THE LIMITS OF THE SELF

THE LIMITS OF THE SELF

Immunology and Biological Identity

Thomas Pradeu

TRANSLATED FROM FRENCH BY

Elizabeth Vitanza

OXFORD
UNIVERSITY PRESS

OXFORD
UNIVERSITY PRESS

Oxford University Press is a department of the University of Oxford. It furthers the University's objective of excellence in research, scholarship, and education by publishing worldwide. Oxford is a registered trade mark of Oxford University Press in the UK and certain other countries.

Published in the United States of America by Oxford University Press
198 Madison Avenue, New York, NY 10016, United States of America.

© Oxford University Press 2012

First issued as an Oxford University Press paperback, 2018

Library of Congress Cataloging-in-Publication Data
Pradeu, Thomas.
The limits of the self : immunology and biological identity / Thomas Pradeu.
p. cm.
ISBN 978–0–19–977528–6 (hardcover); 978–0–19–086957–1 (paperback)
1. Immune system. 2. Immunology—Philosophy. 3. Self (Philosophy) I. Title.
QR181.P64 2012
616.07'9—dc23
2011023261

CONTENTS

Acknowledgments   *vii*

List of Figures   *ix*

Introduction   1

1. Immunology, Self and Nonself   15

2. The Self-Nonself Theory   49

3. Critique of the Self-Nonself Theory   85

4. The Continuity Theory   131

5. Comparing the Continuity Theory to Other
   Immunological Theories   185

6. What Is an Organism? Immunity and the
   Individuality of the Organism   219

Conclusion   267

Notes   *271*

References   *275*

Index   *299*

ACKNOWLEDGMENTS

I would like to thank for their help and support Samuel Alizon, Eric Bapteste, Anouk Barberousse, Cédric Brun, Richard Burian, Edgardo D. Carosella, Stéphane Chauvier, Hannah-Louise Clark, Jacques Dubucs, François Duchesneau, John Dupré, Melinda Fagan, Jean Gayon, Charles Girard, Peter Godfrey-Smith, Alexandre Guay, Philippe Huneman, Richard Lewontin, Marie-Claude Lorne (who was tragically lost in 2008), Michel Morange, Anne Marie Moulin, Maureen O'Malley, Susan Oyama, Jan Sapp, Arthur Silverstein, Kim Sterelny, Alfred Tauber, Mireille Viguier, and Charles Wolfe.

Edgardo D. Carosella, Stéphane Chauvier, Hannah-Louise Clark, Ellen Clarke, François Duchesneau, Jean Gayon, Peter Godfrey-Smith, Richard Lewontin, Michel Morange, Elliott Sober, and Alain Trautmann kindly accepted to read carefully one or several chapters of this book and made very helpful comments. I want to thank them warmly.

In addition, I was very much helped by exchanges with Edwin Cooper, Anthony de Tomaso, Gérard Eberl, Eugene Koonin, Roger Innes, Shimon Sakaguchi, and Lisa Steiner.

Special thanks to Richard Lewontin and Susan Oyama, as well as to Scott Gilbert, in whose ideas I immediately felt "at home." I hope that they will see themselves reflected a bit in this immunological mirror. I also want to thank my former university, Paris-Sorbonne University, the Institut d'Histoire et de Philosophie des Sciences et des Techniques (IHPST) in Paris, and my current employer, the CNRS in Bordeaux, which are extraordinary places for doing research.

Peter Ohlin and Lucy Randall at Oxford University Press have been outstanding editors, always open to my questions or suggestions

And finally, thank you, Magali and Camille.

LIST OF FIGURES

1.1. The human immune system 33

1.2. The blood cells in mammals, with
their main functions 34

1.3. Subfamilies of T cells 35

1.4. The drosophila's humoral immune response 38

1.5. The plant's NBS-LRR immune system 40

3.1. The intensive interactions between normal
bacteria and the gut's immune system 120

4.1. General principle of the continuity theory 138

4.2. Activation of a dendritic cell and a T cell
according to the continuity theory 140

4.3. Phagocytosis according to the continuity theory 164

4.4. The activation of regulatory T cells
according to the continuity theory 167

4.5. The continuity theory unifies under a unique
explanation many immune phenomena 181

INTRODUCTION

Can two individuals that are said to be "identical"—twins, for example—nevertheless be distinguished from one another? Where does a colonial organism, such as a sea coral, begin and end? Is such a colonial organism one or several individuals? What ensures that the larva is the "same" living thing as the adult fly it becomes despite the considerable changes it undergoes? All of these questions form a more general problem: What makes the identity of a living thing? This problem of the definition of biological identity is the one I put forth in this book. Once I have defined the problem, I will show why one discipline among the modern life sciences, immunology, has made the study of this question of biological identity its own domain.

WHAT IS BIOLOGICAL IDENTITY?

The question "What makes X's identity?" can be asked of any entity, even inert objects. Here, the focus is on living things: "What makes the identity of a living thing?" In reality, to ask what makes the identity of a living being is to ask two questions: on the one hand, what makes the *uniqueness* of a living thing, and on the other, what makes its *individuality*. The first question, that of uniqueness, is the following: What makes a living thing different from all other living things, including those that belong to the same species? For example, are there means of distinguishing between two identical twins? Or between an organism and its "double" created through cloning? The second question, that of individuality, is: What counts as *one* living being? In other words, what constitutes a discrete, cohesive, clearly delineated unit in the living world? The problem of individuation is effectively a problem of separation, or delineation, of the real: it consists of knowing how to determine the *boundaries* of the entities being described. It is the problem that is sometimes referred to as that of the "furniture of the world" (what counts as *one* thing, as *one* entity?), applied, in this case, to living things. At least in the domain of the living, an individual is never strictly indivisible—contrary to the etymology of the term "individual." As a result, to understand what creates the unity of a living being consists of determining how it is the unity of a plurality, which is to say why, although it is formed of diverse partially isolatable constituents, the organism is still a unified whole. It is possible, for instance, to wonder what counts as "one" individual in a coral reef: Is a coral one single vast individual whose polyps (each little "tube" topped with a mouth and tentacles) are so many "parts," or should each polyp be considered an "individual"? This example reveals that what biological individuation aims is to offer criteria that allow us to determine precisely what the boundaries of a living being are.

Although they are often confused, uniqueness and individuality are quite distinct from each other: two entities are individuals as soon as it is possible to say that they are two. This does not assume that each entity must necessarily be considered unique. Two tables that would be perfectly identical would not, by definition, be unique; however, they would in fact be two individuals since they could be distinguished and counted as two entities. The same is true of two living things that would be identical while still existing as two separate beings. Thus, the question of biological uniqueness and the question of biological individuality must be distinguished from one another. Of course we could, with Leibniz, affirm that two entities are never completely identical, that they are always unique from a certain number of perspectives. However, in practice, there are entities that we wish to qualify as "identical," particularly in biology: think, for example, of each clone in a clonal plant. The question of biological uniqueness must not then be considered already settled; on the contrary, to answer it requires the establishment of biologically pertinent criteria.

The issue of biological identity, in particular the dimension of biological individuality, is one of the most hotly debated among biologists (Ghiselin 1974; Buss 1987; Maynard-Smith and Szathmary 1995; Michod 1999; Santelices 1999; Queller 2000; Gould 2002; Gardner and Grafen 2009; Queller and Strassmann 2009; Folse and Roughgarden 2010) and philosophers of biology (Hull 1978, 1992; Wilson 1999; Sober 2000; Okasha 2006; Godfrey-Smith 2009). The current literature on levels of individuality and the transitions between these different levels is staggering (e.g., Maynard-Smith and Szathmary 1995; Michod 1999; Queller 2000; Okasha 2006; Godfrey-Smith 2009). The question of biological identity has also been asked by numerous philosophers past and present (Aristotle 1984a, 1984b; Locke [1690] 1979; Leibniz [1765] 1996; Reichenbach [1928] 1958; Strawson 1959; Wiggins 2001). For them, the living being, or rather a certain type of living being, namely an organism, has served as the typical example of what counts as an individual, and, by extension, as the typical example of an entity whose identity can be studied. For Aristotle, in particular, the usual examples of primary substances are the individual man or the individual horse—examples taken up again literally by Wiggins (2001).

The conceptual distinctions that I have proposed are standard in the domain of ontology. However, they are rarely formulated with precision with regard to living beings. My goal is to apply the ontological question "What makes a being's identity?" to the living world, asking "What makes a being's identity in the living world?" The following two questions lie at the heart of this book: First, what counts as an individual in the living world? And second, is each living being unique, and if so what ensures this uniqueness?

One of the branches of contemporary biology, immunology, considers these questions its province. At the heart of immunology, traditionally defined as the science that studies an organism's defense against any foreign entity capable of invading it, have lain the notions of "self" and "nonself" since the 1950s. In setting out to scientifically define the two terms "self" and "nonself," immunologists claim to respond both to the problem of the uniqueness of each living thing and to the problem of its individuality. One of the principal objectives of this book is to establish the precise meaning of these two notions of "self" and "nonself" in order to determine whether they effectively constitute a proper foundation for a definition of biological identity.

THE IDENTITY OF THE LIVING BEING: A CENTRAL QUESTION OF IMMUNOLOGY

From "Self" to Identity

What allows immunologists to lay claim to the problem of biological identity? To answer this question, it is necessary to understand the meaning of the concepts of "self" and "nonself." The self is that which is specific to the organism, that is, both that which defines it and that which uniquely belongs to it. The nonself is everything that is not the self, or what differs from the self's content. For example, in the case of a transplant in an animal, a graft of an organism onto itself ("autograft") is tolerated, whereas a graft from a donor organism onto another ("allograft") is, in almost all cases, rejected. Thus, starting in 1949, Australian virologist Frank Macfarlane Burnet (1899–1985) suggested, drawing specifically on transplantation experiments, to conceive of immunity with the vocabulary of "self" and "nonself" (or "not-self," in Burnet's own words) (Burnet and Fenner 1949). Today, immunologists accept that an organism is capable of an immunological recognition of self and nonself. This distinction between self and nonself allows the organism to trigger a defensive response against all *foreign* entities—which is to say anything different from the self, whereas it will not attack, except for in pathological cases, anything belonging to its self. As Burnet (1941: 60) writes: "There can be little doubt that the whole subject matter of immunology is founded on this intolerance of living matter for foreign matter." As for immunologist Jean Hamburger (1978: 28), he affirms: "Within the same animal species...apart from identical twins, no two individuals are exactly alike....Each individual is able to recognize another individual of the same species as different from himself....Having identified the allografted tissue as foreign, he destroys and eliminates it, while he recognizes fragments of his own body as his own and does not reject them."

What the "nonself" means to the organism, then, is any foreign body that might penetrate it. It can be pathogens (bacteria, viruses, fungi, helminthes, etc.), as well as a graft. In 1949, Burnet and Fenner talked only about "self-markers," but over time the "self-nonself theory" was elaborated to interpret immune reactions from these two central concepts of self and nonself. This theory has dominated immunology for the past sixty years. According to its proponents, the study of the immune system shows that every living being knows its own identity and defends it against any outside threat: it would be able to distinguish between its own components and any foreign one, and would eliminate any foreign body that would penetrate it. Hence, the understanding of biological identity would be immunology's essential and distinctive objective. Indeed, Jan Klein (1982) called it "the science of self/nonself discrimination"; in his Nobel lecture, Jean Dausset (1916–2009) claimed that the system of histocompatibility created the organism's "ID card," which was monitored nonstop by the immune system (Dausset 1981); even Alfred Tauber (2009), who has quite critically analyzed the use of the terms "self" and "nonself" in immunology, argues that immunology is concerned with defining the characteristics of identity that allow one to make the distinction between two individual organisms (along with the question of "describing the mechanisms that defend organisms from their predators"). The recognition of self and nonself by the immune system would therefore make biological identity a question to which immunology could respond. I now turn to how immunology answers, or tries to answer, the twofold question of uniqueness and of individuality in the living being.

Immunology and the Uniqueness of the Living Being

Immunology has appropriated the question of biological uniqueness by integrating and elaborating on the results of genetics. Genetics demonstrates that, in the case of sexual reproduction, two living beings—with the exception of identical twins—are always different, which amounts to saying that every living thing is genetically unique. As for immunology, it furthers this line of research by asking the question of uniqueness at two levels. The first is the immunogenetic level: there is a great diversity in genes involved in immunity, as well as numerous processes of genetic variation and recombination—so much so that, based on a limited number of genes, it is possible to create a huge number of different immune receptors. Moreover, the polymorphism of major histocompatibility complex

(MHC) genes—often called "HLA" for *Human Leukocyte Antigens* in humans—is considerable. The second level is phenotypic: it is expressed in the diversity of immune receptors and the molecules of the histocompatibility system. In mammals, in particular, proteins involved in immunity, especially immunoglobulins, T cell receptors, and MHC molecules, demonstrate an extremely high degree of phenotypic diversity. In humans, for instance, an estimated 5×10^{13} different immunoglobulins and 10^{18} T cell receptors potentially exist. Immune components are therefore one of the most convincing manifestations of each organism's uniqueness, matched only by the phenotypic diversity expressed in the nervous system. In other words, the immune phenotypic characteristics are one of the best ways to distinguish between two individuals or to biologically single out an individual. Dausset (1990: 27) thus says of the HLA system: "The HLA system is the best definition of the being in relation to another individual of the same species, since the experience of transplantation shows us that it is the greatest barrier."

One very important aspect of the phenotypic expression of uniqueness is the construction of a being's uniqueness over the course of time: immune receptors of B and T cells are produced in relation to antigens the organism encounters by virtue of the "immune memory" mechanism. By this mechanism, an organism that reacts to an antigen, for example a bacterium, produces immune receptors specific to that bacterium, then maintains these receptors for its entire life, and will respond more quickly and more efficiently should the same antigen ever be reintroduced. The immune system is consequently said to make an important contribution to the diachronic uniqueness of the living being just as the nervous system does: my immune "self" makes me a unique individual with regard to all other living beings, including those of my own species. Indeed, even two monozygotic twins are different from the perspective of their immune systems.

Immunology and Individuality

Immunology attempts to clarify the question of a living being's individuality by showing that its boundaries are drawn by its immune system, which constantly monitors its components and eliminates everything that is different from itself. It follows that the immune system would be based on the knowledge of self-components and would allow maintenance of the living being's identity by the rejection of any exogenous or foreign entity, in particular pathogens. The immune system would define each living being's individuality and would guarantee the maintenance of its identity over time, what one may call, with Burnet (1962), the "maintaining of integrity." The immune system would assure that *one* being is dealt with and would preserve the unity of this being through time.

The dimension of individuality converges with the dimension of uniqueness in the developments of the transplantation field in the first half of the twentieth century. In showing the acceptance of autografts and grafts from identical twins, transplantation allowed for a precise definition of the expression of individuality in organisms, making this question a fundamental issue of immunology (Loeb 1930, 1945; Medawar 1957; Hamburger 1976).

CONFLICT BETWEEN IMMUNOLOGY AND OTHER BIOLOGICAL FIELDS ON THE QUESTION OF BIOLOGICAL IDENTITY

Immunologists therefore rely on apparently solid experimental arguments when they claim that they are dealing with the identity of the living being in the double sense of its uniqueness and its individuality. It appears highly unlikely, however, that immunology could be the *only* biological discipline capable of tackling this problem. For instance, the question of uniqueness seems to pertain much more to genetics than to immunology: the uniqueness of sexually reproducing organisms is first and foremost genetic in origin. Immunologists certainly put forth immunogenetic and phenotypic arguments in order to say that each living being's uniqueness is even stronger than that which can be ascertained by genetics, but in these circumstances it would be much exaggerated for them to claim the question of biological uniqueness as theirs alone.

As for the question of individuality, if it has been at the center of numerous biological and philosophical debates for the past thirty years, it has been so almost exclusively from the perspective of evolution (Ghiselin 1974; Hull 1978; Buss 1987; Maynard-Smith and Szathmary 1995; Michod 1999; Santelices 1999; Queller 2000; Gould 2002; Okasha 2006; Gardner and Grafen 2009; Godfrey-Smith 2009; Queller and Strassmann 2009; Folse and Roughgarden 2010). The theory of evolution by natural selection provides a response to the question of biological individuation by defining a hierarchy of "evolutionary individuals," that is, entities upon which natural selection acts (Lewontin 1970; Hull 1978; Buss 1987; Gould 2002; Okasha 2006; Godfrey-Smith 2009). In this hierarchy, the organism appears only as one of several possible biological individuals, along with the gene, the genome, the cell, or even the group, the species, etc. As a result, it seems difficult for immunology to consider that it alone can respond to the question of biological individuality.

Immunology must therefore take into account other biological disciplines that also claim to clarify the question of biological identity. It will be critical, in this book, to determine what relation immunological discourse supports with these other disciplines:

Are they complementary or conflicting approaches? Or do they perhaps envision different objects, and do not mean the same thing when they resort to the terms "uniqueness" and "individuality"?

THE PROBLEM OF SCALE WHEN DETERMINING BIOLOGICAL IDENTITY

The confrontation that I have just briefly sketched out between immunology and evolutionary biology on the question of individuation brings up another important concern: What is the scope of this notion of biological identity? In effect, individuation as explained by the theory of evolution appears to apply to the entire hierarchy of living beings (genes, cells, organisms, etc.), whereas individuation as explained by immunology only seems to concern itself at the level of the organism. Yet nothing indicates that the question of the living being's identity must be asked exclusively at the level of the organism.

Even if we admitted that the problem of biological identity arises at the level of an organism and not at other biological levels, a crucial issue would remain: With exactly which organisms is immunology concerned? In order for immunology to offer a general conception of what makes the identity of an individual organism, it would be necessary for it to apply to all organisms, or at least to the great majority of them. Now, for Burnet, the founder of the self-nonself theory, immunology applies exclusively to higher vertebrates, as he thought that they only had a true "immune system." If Burnet is right, therefore, immunology cannot claim to shed light on the question of a living being's identity except for a tiny fraction of living things. This would put immunology at a considerable disadvantage compared with evolutionary biology, which deals with numerous levels of life.

BIOLOGY'S QUEST FOR IDENTITY

The preceding analyses thus raise the following questions:

1. Does the immunological self-nonself theory allow for the elaboration of a satisfying conception of biological identity? I will show that this theory is inadequate from an experimental point of view and imprecise from a conceptual point of view. I will conclude that this theory cannot lead to a satisfying conception of biological identity.

2. Does another immunological theory allow for the creation of a satisfying conception of biological identity? I will propose another theory, the "continuity theory," by highlighting the experimental arguments that appear to validate it. I will show that this theory permits the elaboration of a satisfying conception of biological identity.

3. What precise meaning of the term "identity" applied to living beings can be clarified by immunology? I will point out that immunology brings precise and decisive elements to bear on *one* particular meaning of the term, that of diachronic individuality in living things.

4. On what scale of life does the immunological conception of biological identity belong? I will show that it applies only at the level of organisms, but that it does include all organisms from unicellulars to multicellulars. This scope is broad enough to clearly show immunology's usefulness in dealing with the question of biological identity.

5. Where does using immunology to respond to the question of biological identity lead? I see it as leading to a precise definition of what counts as an organism. One of the theses I shall defend is that the organism is not an endogenous reality; in other words, the organism is not a set of

biological components originating from successive autonomous divisions of an egg cell. I will show why and in what sense the organism is, on the contrary, a heterogeneous reality made up of components of different origins.

6. How can the contribution of immunology and that of other fields, in particular that of evolutionary biology, be articulated in order to fully understand biological identity? I will argue that immunology allows a physiological individuation of living beings that, once it is joined to the evolutionary individuation of living beings, helps to clarify the latter and shows that the organism is probably the most highly individuated biological entity.

Thus, my overarching question is to know what makes the identity of a living being. To attempt to respond to this inquiry, I will examine the current state of immunology. This will be partly historical, centered particularly on the theses Burnet proposed from the 1950s to the 1970s. Nevertheless, my analysis will mainly consist of a conceptual and theoretical investigation into today's immunology, as well as a scientifically grounded examination of an ontological question. My objective is to show that the understanding of biological identity cannot overlook results of current immunology. In other words, immunology is essential to the question of the identity of living things. Indeed, I hope to show that immunology offers an original and fertile starting point for defining biological identity in its modality of what makes a living thing's individuality over time.

Six chapters constitute this work. In the first, I define immunology and two of its fundamental concepts—those of the self and nonself. In the second, I analyze the self-nonself theory that dominates current immunology. In the third chapter, I offer a critique of this theory. In the fourth, I propose my own theory, the continuity theory, by showing what it is based on and which solutions it provides to the challenges raised in the third chapter. In the fifth chapter, I draw a comparison between the continuity theory and other available immunological theories. Finally, in the last chapter I show how the concept of immunity that I am arguing for offers a definition of the organism as a biological individual.

The double meaning of this work's title should now be quite obvious. To speak of the "limits of the self" comes down to the demonstration of two ideas: first, the insufficiency of the self-nonself theory, and second, the possibility for immunology to respond to the decisive question of delineating the frontiers of the organism— understood from this point forward as a "self" to be redefined.

Chapter 1

Immunology, Self and Nonself

In this chapter, I present a definition of immunology, and I defend the view that all organisms have an immune system. To provide the reader with clear, accessible information, I detail the way the immune systems of organisms belonging to different species work. Finally, I explain why immunology has been considered as the "science of self and nonself" and what the different meanings of the word "self" are in immunology.

WHAT IS IMMUNOLOGY?

The Traditional Definition of Immunology: Immunity as Defense

Immunology is the science of immunity. The word "immunity" comes from the Latin *immunitas*, "exemption, deferral," derived from *immunis*, which, in Roman law, meant "exempt from all charges or taxes." The word's origins therefore refer to the idea of exception, of specificity. By extension, biological immunity is considered to be an organism's capacity to react to a pathogen and so escape its pathology. Immunology thus appears to be the study of the means by which an organism escapes the pathogenic effects of certain substances.

Immunology has traditionally been defined as the discipline that studies *defense* systems of living things against pathogenic entities, which is to say those that can induce illness, whether it be viruses, bacteria, archaea, fungi, etc. Consequently, the immune system is often defined in warlike terms to underscore the existence of a struggle between "them" (the pathogens) and "us" (the organism defending itself). Immunological vocabulary is loaded with words that have connotations of war, as in "natural killer cells" or "cytolytic T cells." I will show later on that the immune system is more a system of regulation that involves endogenous and exogenous entities alike than a defense system against exogenous entities. Nevertheless, this definition of immunity as a defense system has dominated immunology during most of its history, and it remains the most commonly used definition, both by immunologists (e.g., Janeway 2001; Clark 2008) and by the general public.

The Formation of Immunology as an Independent Discipline

The question "what is immunology?" can also be answered from an institutional point of view; however, articulating what is unique to immunology with regard to other biological fields is quite difficult. Before the 1930s, immunology did not constitute a specific discipline and was not studied as such by doctors or biologists. I would like to distinguish three main phases in the formation of immunology as an independent science: immunization; elaboration of a theory of immunity; and finally positioning of immunological discipline per se.

The first phase in forming the independent field of immunology was that of *immunization* or scientific vaccination. Immunization is the process by which an organism's contact with a pathogenic agent allows the organism to ultimately resist the pathogen's destructive effects. Determining the historical origins of vaccination is difficult, since traces of it are found in myriad places around the globe, from tenth-century China to the Ottoman Empire of the eighteenth century, without it being easy to specify one widely accepted momentous first occasion of immunization (Moulin 1991).

It is only in the nineteenth century, however, that immunization becomes "scientific," in the sense that it is performed on a wide scale and leads to the eradication of many illnesses. The three major figures in this era are Englishman Edward Jenner (1749–1823), the German Robert Koch (1843–1910), and the French Louis Pasteur (1822–1895). Jenner is at the origin of "vaccination" in the sense of the inoculation of man with the cowpox virus ("vaccine") in order to prevent smallpox. Robert Koch's contribution is of major importance: he and his team show that microorganisms cause infectious diseases and that each one of these is responsible for a particular pathology. As far as Pasteur is concerned, he uses the term "vaccination," though it is meant to refer generally to all inoculation of weakened or inactive germs by laboratory procedures to prevent the breakout of illness. Pasteur's era is marked by great ambitions and hopes concerning the possibility of eliminating all illnesses. It remains the most important time period for understanding what exactly reflections on immunization brought to the formation of immunology (Janeway 2001; for a critical eye, see Grmek 1996). Immunization is an important step because it brings awareness of the existence of what comes to be called immune "memory." This term describes an organism's capacity to react in a faster and more intense way upon a second contact with a given antigen (as will be detailed later in this chapter, the term "antigen" is broader than the term "pathogen": an antigen is any substance that can trigger an immune response). The immune system appears capable of "learning" to attack an antigen more efficiently. It is this phenomenon of immune memory that is used in the case of vaccines. At the same time, immune memory is not a necessary component of immunity.

Contemporary immunology distinguishes two types of immunity, one said to be "innate," the other "adaptive." Organisms endowed with adaptive immunity—in other words, those with an immune "memory"—are those that are capable of triggering a faster and more effective immune reaction in the case of a second encounter with an antigen.

The second phase in the formation of immunology corresponds to the elaboration of a theory of immunity. This is owed to the Russian zoologist Elie Metchnikoff (Tauber and Chernyak 1991). Metchnikoff proposed placing phagocytic cells or macrophages at the center of immune functioning. According to him, these cells are responsible for all forms of pathogen destruction and of maintenance of the organism's cellular equilibrium (phagocytes—from the ancient Greek *fagein*, to eat—swallow both pathogens and the organism's dead cells, as well as inert elements). Metchnikoff's theory is the "cellular theory of immunity." It enters into conflux with the "humoral" theory of immunity, represented, in particular, by Robert Koch and Emil von Behring (Silverstein 1989: 38–58). The humoral theory states that the active immune agents are soluble components (molecules) found in the organism's "humors," and not cells. The struggle between these two theories dealing specifically with immunity constitutes a decisive step in the formation of immunology as a unified, autonomous discipline.

The third phase is the discipline's *institutionalization*. It is only in the 1930s that immunology begins to be considered as an independent discipline, identified as a specialty in university courses of study, first for doctors, and then for biologists. Up until then, it was impossible to speak of immunology as a well-defined field within biological sciences as it was, most of the time, considered as a branch of microbiology. The first academic chair dedicated solely to immunology seems to have been that of Arthur Coca at Cornell University in 1925 (Moulin 1991: 121, 140). Yet even after the creation of this first chair, followed slowly by others around the world, immunology only played, until the 1960s, a secondary role in university curricula. The 1960s constitute a decisive turning point for immunology, which becomes independent by developing into one of the most dynamic branches of biology—which it still is today. Experiments on human transplantation have played a major role in immunology becoming an independent and influential discipline (Mazumdar 1995). These experiments also contributed, via the idea of donor "compatibility," to the elaboration of the concept that immunology's goal is the "defense against nonself," the "nonself" being all that is foreign to the organism.

The Definition of Immunology Proposed Here

The definition of immunology as the science studying organisms' defense against pathogens is strongly linked to the self-nonself theory. Throughout the rest of this work, I will critique the validity of the self-nonself theory. Hence the following question: Does my rejection of the self-nonself theory leave me without a definition of what I mean when I talk about immunology? On the contrary, I offer here a structuring definition of immunology as *the discipline that studies specific interactions between immune receptors and antigenic patterns (ligands), interactions that can trigger mechanisms that destroy or prevent the destruction of target antigens.* I think all immunologists, including those who defend the self-nonself theory, can accept this definition. Let me now explain in detail what this definition means.

According to this definition, immunity is the capacity to escape antigens in a specific way. I must then begin by defining what an antigen is. Originally, an *antigen* was any substance capable of binding itself to an *antibody* (*immunoglobulins* are surface molecules of B cells; they are called *antibodies* when they are secreted). Today, the term *antigen* generally refers to any molecule (ligand) capable of setting off an immune reaction in an organism. However, every *antigen* is not necessarily an *immunogen*: only antigens that effectively trigger an active immune response are called immunogens. In addition, every *antigen* is not necessarily a *pathogen*: only antigens that trigger damages or disease in the host are said to be pathogens. And finally, as will become clearer in the rest of this book, an antigen is not necessarily *foreign*.

According to this definition, there is immunity, properly speaking, only when there is a biochemically specific reaction between an antigen (ligand) and receptors carried by the immune system's actors. Consequently, in my conception, the traditional distinction between "specific" and "nonspecific" immunity loses its relevance: normally, "specific" immunity is characterized by the great diversity of receptors carried by the cells involved, and therefore refers essentially to immunity provided by B and T cells, whereas "nonspecific" immunity refers to immunity provided by epithelial surfaces, macrophages, dendritic cells, the complement, natural killer (NK) cells, etc. Yet in the definition that I am proposing, this dichotomy is completely modified. Certain components of immunity said to be "nonspecific" will be considered quite simply beyond the field of immunity, since no specific biochemical interaction with the antigen is produced by them (examples include epithelial surfaces as such, as well as undifferentiated antifungal substances), whereas other so-called "nonspecific" actors of immunity will actually be classified within immunity (always specific, following my definition), in the sense that they do use receptors capable of reacting with antigens, in which case their alleged "nonspecificity" actually means that the pattern they recognize specifically is widespread in nature. A good example is Toll-like receptors (TLR) located on antigen-presenting cells (APCs). These surface molecules, named after "Toll" receptors discovered in fruit flies, are capable of binding to widespread pathogenic patterns (Medzhitov 2001). I therefore include in my definition of immunology macrophages, dendritic cells, NK cells, etc.—i.e., any immune cell with receptors. I would like to stress here that it is important to not confuse the absence of biochemical specificity with biochemical specificity for an often-repeated pattern (Vivier and Malissen 2005). Any immune response presupposes a specific reaction between an antigen and receptors carried by immune actors.

My definition of immunity depends therefore on two arguments: (1) There is immunity only when there is a specific reaction with an antigen (i.e., all immunity is specific); (2) there is immunity when there is a receptor capable of interacting with ("recognizing") an antigenic pattern (ligand), even if this antigenic pattern is repeated in nature.

Immunology is thus the study of all the specific binding reactions between the organism's immune receptors and antigenic patterns (ligands). At this point, an objection might be raised: Isn't my definition circular, since "immunology" is defined by the term "immune receptors"? It is not circular, because the different types of immune cells, as well as the binding reactions between cellular receptors and antigens, are clearly defined and observable. Consequently, one can establish a list of cells whose receptors can bind to antigens and define immunology as the study of all of these specific reactions. In mammals, these cells are monocytes, which differentiate into macrophages, mast cells, dendritic cells, granulocytes or polymorphonuclear leukocytes (neutrophils, eosinophils, basophils), lymphocytes (B and T cells), as well as natural killer cells (more will be said on all these cells in the section "Presentation of the immune system" below). I thus define immunology by its objects, which is often the case when one tries to define a science. The definition offered here is a biochemical definition, since the criterion adopted here is the biochemical specificity of interactions between receptors and antigenic patterns (ligands). But how do we know which immune cells should be included and which should be left out? As I stated in my preliminary definition, what defines these cells is their reactive capacity, i.e., their ability to trigger activating mechanisms (either effector or inhibitory mechanisms), and, by extension, their ability to destroy or prevent the destruction of entities with which they react. Consequently, since the specific interactions between immune receptors and their ligands can be observed and quantified, the proposed definition is not circular.

WHICH ORGANISMS HAVE AN IMMUNE SYSTEM?

Traditional and More Recent Views about the Scope of Immunology

The definition of immunity I have just given helps me clarify a decisive point, one which I will return to constantly in the analyses that follow: What are the organisms that possess an immune system? Given that this is a crucial problem, I am going to introduce Burnet's (the immunologist who has had the most influence on the discipline) answer to this problem, before showing why I am going to adopt a broader concept of the field of immunology, in claiming that all organisms—unicellulars and multicellulars—have an immune system.

According to Burnet, only jawed vertebrates possessed a true immune system. Without getting too far ahead into the following chapter, which will show in depth the scope of the self-nonself theory, I would here like to illustrate two points that explain why Burnet restricted the scope of immunity to jawed vertebrates. First, Burnet thought that only the immunity provided by lymphocytes was effectively specific, and thus constituted a genuine immunity. Second, in a vision of immunology that was marked by the significance of vaccination phenomena, he considered that actual immunity existed only where there was immune memory (until very recently there was a consensus that, in the course of evolution, it is only with jawed vertebrates that immune memory appeared). These two reasons explain in large part why, in Burnet's thinking, immunity in the strict sense could only concern higher vertebrates. Although in the next chapter I will better explain his immunological theory, it is sufficient for now to point out how contemporary immunology has modified Burnet's conception of the extension of the field of immunology.

Most immunologists today consider, contrary to Burnet, that it is possible to speak of immune systems in all animals, vertebrates and invertebrates alike. Invertebrates do possess immune systems, which are often as effective and complex as those of vertebrates. The idea that invertebrate animals have an immune system was expressed long ago (e.g., Cooper 1974), but it became widely accepted only with the accumulation of staggering data by researchers working on what is referred to as "innate" immunity (Janeway 1989; Janeway and Medzhitov 2002). Work on insects, most prominently fruit flies, played a decisive role in demonstrating the existence of an immune system in invertebrates (e.g., Lemaitre et al. 1996; Lemaitre, Reichhart, and Hoffmann 1997). We can talk of a true immune system in fruit flies because, besides the epithelia that make up physical barriers to microbe entry, the fruit fly has several immune mechanisms at its disposal (Lemaitre and Hoffmann 2007; see also the detailed description of the fruit fly's immune system and the accompanying figure 1.4 below). These mechanisms are mainly twofold: a cellular response, the phagocytosis of pathogens, which is carried out by cells called "plasmatocytes"; and a humoral response, provided by the "fat body" (the equivalent of the mammalian liver). These mechanisms involve genuine specific interactions with surface patterns carried by the targets (Janeway and Medzhitov 2002;

Brennan and Anderson 2004). The cellular response is principally achieved by plasmatocytes, which can eliminate microorganisms, as well as apoptotic cells—those going under programmed cell death. It relies on an interaction between specific receptors and patterns situated on the surface of microorganisms or apoptotic cells (Ramet et al. 2001; Kocks et al. 2005). The humoral response mobilizes immune receptors that belong to several families, the best known of which is that of "Toll" receptors. Contrary to what occurs in the case of the cellular response, the recognition that triggers the humoral response is said to be "indirect" because it is carried out by distinct proteins rather than by immune cells' surface receptors. Indeed, the detection of microbes requires direct contact between a host's protein called *pattern recognition receptor* (PRR) and a microbial molecule. This involves two main families of proteins, *peptidoglycan recognition proteins* or PGRP (Michel et al. 2001) and *Gram-negative binding proteins*. Following the interaction with pathogens, genes coding for specific antimicrobial peptides are activated. The fruit fly's complex immunity rests on specific recognition pathways and leads to the destruction of a great variety of pathogens. It also displays numerous common points with vertebrate immunity, reflecting, in some cases, the preservation of certain mechanisms in the course of evolution (Khush, Leulier, and Lemaitre 2002).

Another decisive element illustrating invertebrate immunity came from the demonstration that, counter to what had been believed for decades, several invertebrates possess "immune memory" mechanisms, meaning that they can trigger a faster and more effective immune response upon a second exposure to an antigen. The existence of mechanisms of increased immune response intensity and speed was effectively shown in invertebrates (Kurtz and Franz 2003; Hemmrich et al. 2007; Litman and Cooper 2007). Many invertebrates therefore possess adaptive immunity (Alder et al. 2005; Kurtz and Armitage 2006; Litman and Cooper 2007), a fact that eliminated what had long been considered one of jawed vertebrates' exclusive properties (Lanier and Sun 2009; Pradeu 2009). Moreover, gene rearrangements that create a highly diverse immune repertoire do occur in jawless vertebrates—against, here again, what had long been thought. It seems that this jawless vertebrates recognition system, based on *variable lymphocyte receptors* (VLR) evolved separately from that of jawed vertebrates (Pancer et al. 2004; Saha, Smith, and Amemiya 2010).

The thesis that I defend here, and which will ultimately allow me to propose a truly general theory of immunity, is that immune systems do exist in all organisms, both unicellular and multicellular. Indeed, contemporary immunologists discovered specific interaction mechanisms capable of leading to the destruction of a target in every organism they researched. In other words, everywhere in the living world, and opposed to what was believed just ten or fifteen years ago, immune mechanisms are at work. In one sense, that should not be surprising, since all organisms, even the simplest, are subjected to pathogen actions (Cooper 2010a). Yet the question that follows is whether these organisms possess *specific* recognition mechanisms for pathogens. As I stated previously, the idea that insect immunity exists is accepted by every immunologist today, after having first raised numerous objections. I will now explain why it is possible to extend the scope of immunity even further—to plants, and even to unicellular organisms.

The Existence of Plant Immunity

Plants possess numerous immune mechanisms (Taiz and Zeiger 2006), including extremely specific ones. One such mechanism consists of the recognition of *pathogen-associated molecular patterns* or PAMPs (Chisholm et al. 2006). This immunity is induced by the recognition, thanks to receptors located on the plant's cell surfaces, of microbial patterns conserved in the course of evolution (Nurnberger et al. 2004). For instance, plants recognize multiple surface components of Gram-negative bacteria, including lipopolysaccharide and flagellin. This recognition of microbial patterns takes place due to proteins found on the cell surface of the plant, as, for example, in the case of a flagellin receptor in the *Arabidopsis* plant, a kinase called *"receptor-like kinase,"* that consists of *leucine-rich repeats* (LRR) and an intracellular serine/threonine domain. One of the responses the plant may trigger is what is called a *hypersensitive response,* which is a form of programmed cell death localized at the level of the infection's site. This local, limited response is the result of a specific interaction with the pathogen, allowing the plant to circumscribe the place where it releases the programmed cell death.

A second immune mechanism in plants consists of the specific recognition of pathogenic effectors. Pathogens, whether bacteria, viruses, or fungi, secrete countless effectors that inhibit or at least impede the first immune mechanism I described, PAMP recognition. The second mechanism, known for over thirty years and called *gene for gene resistance* (Flor 1971), allows plants to interact in a specific way with these pathogenic effectors (also sometimes called "Avr," for *avirulence proteins*) and to eliminate them. This resistance leads to the induction of programmed cell death at the site of infection and to the inhibition of the pathogen's growth. Plants possess resistance genes (called "R") that code for proteins that recognize specialized pathogenic effectors. From a biochemical perspective, the application of the "gene for gene" hypothesis is a receptor-ligand model in which plants activate defense mechanisms after the recognition, thanks to R proteins, of pathogenic products coming from the pathogen (Van der Biezen and Jones 1998; Boller and He 2009). Plant resistance proteins fall into two categories. The first category is a family of proteins containing a *nucleotide binding site*

(NBS) and *leucine-rich repeat* (LRR) domains. In *Arabidopsis*, there are more than 150 proteins from the NBS-LRR family (Dangl and Jones 2001). NBS-LRR proteins also exist in animals, where they make up an important part of the innate immune system (Rairdan and Moffett 2007). The second category comprises proteins with *extracellular leucine-rich repeat* ("eLRR") domains. Together, these proteins take part in resistance to all types of pathogens.

Plants can thus interact specifically with antigenic patterns, and these interactions can lead to the target's destruction, following a receptor-ligand model. To speak of immunity in plants is therefore possible, in the precise sense of the term I have proposed here. In particular, the second mechanism described above, the one that involves specific recognition of pathogenic effectors, shares several important aspects with vertebrate adaptive immunity, especially mammal adaptive immunity (DeYoung and Innes 2006: 1243).

Finally, in numerous plants one can observe a *systemic acquired resistance* phenomenon that might partly resemble an "immune memory" phenomenon but which is actually nonspecific, with a limited duration of a few hours to a few days (Ryals et al. 1996).

The Genome's Immunity? The Different Forms of "RNA Silencing"

Increasing numbers of immunologists agree that, in addition to the organism's immunity, there is a genome's immunity (Plasterk 2002; Ding 2010). This genetic immunity bears the generic name "RNA silencing." RNA silencing occurs in all eukaryotic organisms, i.e., those with cells that have a nucleus. RNA silencing appears to satisfy the definition of immunity I have put forth. In any case, if in the years to come immunologists confirm the thesis that this is truly a form of immunity, the field of immunology will have been considerably broadened.

The term "RNA silencing" refers to different mechanisms that all share the same functioning principle: the silencing of a specific gene by small silencing RNAs (a generic class that presently comprises *small interfering RNAs, microRNAs,* and *Piwi-interacting-RNAs,* but this list will certainly be extended in the near future) (Ding 2010). The most important and best characterized RNA silencing mechanism is "RNA interference," which originally referred to the inhibition of the expression of a specific gene, after double-stranded RNA activates a biochemical complex that degrades messenger RNAs bearing the same genetic code as the double-stranded RNA (today, "RNA interference" refers to specific gene silencing by both double-stranded RNA or small interfering RNAs). Andrew Fire and Craig Mello discovered RNA interference in the *Caenorhabditis elegans* nematode in 1998 (Fire et al. 1998), which earned them the Nobel Prize in Physiology and Medicine in 2006. The two researchers showed that the synthesis of certain proteins could be prevented in a targeted way by introducing a double-stranded RNA into the nematode's cells. The mechanism of RNA interference is as follows: double-stranded RNA binds to a protein called Dicer, which cleaves double-stranded RNA into fragments; one of the two RNA strands is loaded into a complex called RISC (*RNA-induced silencing complex*), while the other strand is eliminated; when a messenger RNA (mRNA) pairs with the RNA fragment on RISC, it is degraded, which results in the inactivation of the corresponding gene.

It quickly became clear that RNA silencing was an immune mechanism located at the level of the genome, since it allowed for the elimination of pathogenic genomes. RNA silencing is therefore an essential, ubiquitous form of immunity (found in plants, fungi, algae, invertebrates, vertebrates) (Ding 2010). For instance, RNA interference is the fruit fly's only defense mechanism against viruses to be identified today, while this species must surely face a robust viral threat (Wang et al. 2006; Lemaitre and Hoffmann 2007; Saleh et al. 2009).

I think that my definition of immunology applies to RNA silencing in general, and to RNA interference in particular. Indeed, in these phenomena, there is specific recognition of a target RNA, followed by an effector response. This recognition involves specific nucleic acids; that is to say, there is certainly a change of scale with regard to the usual immune response (we move from recognition mainly performed by proteins on the cell's surface to intracellular recognition), but the nature of the phenomenon is the same. Moreover, the mechanism of RNA silencing does not function solely against "foreign" RNAs; instead it seems to work on abnormal RNAs whether the latter are endogenous or exogenous (Sontheimer and Carthew 2005).

In addition, it is important to note that strong links exist between the cellular level of the immune response and the genetic level. Thus, a number of innate immune receptors, notably certain Toll-like receptors (TLR) such as TLR9 (Ishii and Akira 2006), detect pathogens inside the cell (Lee and Iwasaki 2007), at the DNA and RNA level. In general, the past ten years have brought to light an entire surveillance system unique to the genome.

The preceding analysis on the genome's immunity leads to an important question: Do unicellular organisms like bacteria have genetic immunity, or more generally, mechanisms that can be qualified as "immune"? The answer to this question is difficult and complex. I will focus my analyses here on bacteria and archae.

The Existence of Immunity in Single-Celled Organisms

Bacteria and archae are under the pressure of pathogens, first and foremost viruses called "bacteriophages" (Burnet 1934). Bacteriophages make up the largest group of all known viruses and are arguably the most abundant biological entity on the planet (Chibani-Chennoufi 2004). In addition, they are one of the simplest biological entities that exist. Bacteria have developed response mechanisms to these commonly occurring pathogens. In the course of the evolutionary "arms race," bacteria acquired the ability to identify and to eliminate pathogenic agents that penetrated them. The question is how exactly this identification occurs.

Studies on bacteria used in industrial milk fermentation shed important light on bacteria's natural resistance mechanisms to bacteriophages. These mechanisms range from the blocking of phage DNA injection to abortive infection systems interfering with phage DNA replication, RNA transcription, phage development, and morphogenesis (Chibani-Chennoufi et al. 2004).

Can we speak of a true "immune system" in these bacteria? Many researchers think so. Among them are Kira Makarova, Eugene Koonin and colleagues, who have recently shown evidence in bacteria of a mechanism analogous to eukaryotic RNA interference (Makarova et al. 2006). They claim that CRISPR (*clustered regularly interspaced short palindromic repeats*) and *Cas* (*CRISPR-associated*) genes play the decisive role in this mechanism. CRISPRs are a class of repetitive elements that are present in numerous prokaryotic genomes. A CRISPR element consists of a direct repeat of 28 to 40 base pairs, with the copies separated by a unique sequence of 25 to 40 base pairs. Koonin and his colleagues propose the hypothesis that "the CRISPR-Cas system is a mechanism of defense against invading phages and plasmids that functions analogously to the eukaryotic RNA interference systems." Koonin and his colleagues' claim rests mainly on two observations. First, the fact that at least certain unique CRISPR inserts come from phage or plasmid genes. Second, the abundance of components associated with the CRISPR system that are clearly involved in nucleic acid deterioration, presentation, and perhaps recombination.

The hypothesis of Koonin and his colleagues has recently received an important confirmation (Barrangou et al. 2007). The authors of this article, in an analysis of the CRISPR sequences of several lines of *Streptococcus thermophilus*, show first that, in becoming resistant to bacteriophages, the CRISPR1 locus of bacteria is modified by the integration of new *spacers*, apparently derived from the DNA of the bacteriophage itself; and second, that the presence of a CRISPR *spacer* identical to the phage's sequence brings resistance against phages containing this particular sequence. The authors conclude that "prokaryotes appear to have evolved a nucleic acid–based 'immunity' system whereby specificity is dictated by the CRISPR spacer content, while the resistance is provided by the Cas enzymatic machinery," and talk about an "adaptive" bacterial immunity (the same idea is expressed by van der Oost et al. 2009).

In a recent paper, Horvath and Barrangou suggest that CRISPR interference is in fact not mechanistically analogous to eukaryotic RNA interference, but rather is definitely an interference-based adaptive immune system, to be found in 90% of archaea and 40% of bacteria. This system makes possible an immune "memory" (acquired immunity) that can be transmitted to offspring (Horvath and Barrangou 2010).

I take from these analyses that archaea, bacteria, and more generally unicellular organisms, most certainly possess an immune system in the sense of a group of molecular mechanisms that allow specific interactions with pathogenic patterns (especially viral patterns) and the elimination and inactivation of these pathogens.

Although I have tried to present the immune mechanisms of fruit flies, plants, and single-celled organisms in the most intelligible way, it is not easy to describe such complex phenomena in simple terms. I hope, however, that the conclusion drawn from this analysis is, for its part, perfectly clear: immunity is a ubiquitous phenomenon in nature.

In particular, I think that it is highly likely that further studies will confirm that it is possible, for unicellular organisms, to speak of an "immunity" in the sense I have defined. That said, even if it appeared that my definition actually only applies to multicellular organisms, that would already be, if I am not mistaken, a remarkable result, since such a definition would in itself offer a much broader perspective on immunity than that of Burnet. Such an extension will allow me to propose a truly general theory of immunity. Throughout this book, then, when I refer to immunity, it is as much with plants and insects, and even unicellular organisms in mind, as with human beings.

PRESENTATION OF THE "IMMUNE SYSTEM"

Starting from the definition of immunology that I have proposed, I now turn to a quick description of the group of components usually referred to as the immune "system." I will only discuss here briefly the human immune system before moving on to fruit fly and finally plant immune systems.

The Human Immune System

The human immune system is well known because of its obvious medical interest. Nevertheless, it is important to underscore that the immune systems of most mammals, particularly mice, are very similar, and that what I describe here with regard to the human immune system thus has a certain level of generality.

The human immune system comprises a group of specific organs, including the bone marrow, the thymus, the spleen, the lymph nodes; and a circulation system, namely the lymphatic system, which communicates with the blood system (see figure 1.1). The bone marrow and the thymus are called the "primary lymphoid

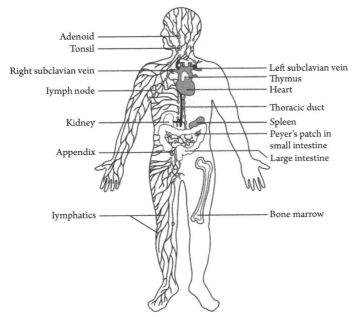

Figure 1.1. The human immune system: The immune system exerts its influence through-out the organism. In mammals, it is made of lymphatic vessels and key organs such as the bone marrow, the thymus, the spleen, the lymph nodes, the Peyer's patches, etc. (Copyright from Janeway et al. 2005)

organs," because they are where the lymphocytes are produced; the lymph nodes, the spleen, and the Peyer's patches in the intestine constitute the "secondary lymphoid organs," because they are where lymphocytes interact with their specific antigens.

The human immune system also comprises different cells (see figure 1.2), including monocytes, which, when they migrate to tissues, differentiate into macrophages (the organism's "garbage men"); dendritic cells (which are "professional" antigen-presenting cells); lymphocytes (B and T cells); NK cells; mast cells; granulo-cytes or polymorphonuclear leukocytes (divided into three catego-ries: neutrophils, eosinophils, basophils). Finally, several molecules,

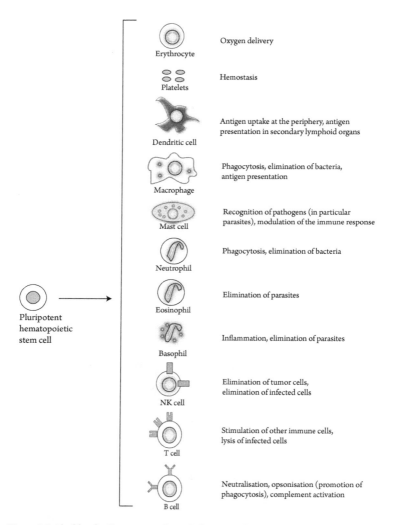

Figure 1.2. The blood cells in mammals, with their main functions: All immune and blood cells originate from a unique pluripotent hematopoietic stem cell. Immune cells are sometimes called "leukocytes" or "white blood cells." It is usually considered that "innate" immune cells encompass dendritic cells, macrophages, mast cells, granulocytes or polymorphonuclear leukocytes (that is, neutrophils, eosinophils, and basophils), and natural killer (NK) cells, while "adaptive" immune cells would encompass T and B cells. These distinctions have nevertheless tended to blur in recent years.

mainly cytokines, but also the "complement" system, play a decisive role in the immune system. All of this system's elements ensure the specific recognition by which I have defined immunology.

This presentation is purposely oversimplified. As a matter of fact, many of the cellular categories just mentioned further divide into several subfamilies. An important example is that of T cells. As figure 1.3 shows, T cells divide into at least five subfamilies. *Helper* T cells (which usually express the CD4 surface molecule) stimulate other immune components. For the sake of clarity, only three

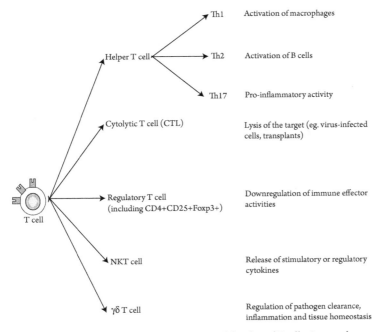

Figure 1.3. Subfamilies of T cells: There are many subfamilies of T cells. Among the most important are helper T cells, which activate other immune components; cytolytic T cells, which induce the destruction of their target; regulatory T cells, which dampen the activity of effector immune mechanisms; natural killer T (NKT) cells, which release stimulatory and regulatory cytokines; and γδ T cells, which are involved in the regulation of pathogen clearance, inflammation, and tissue homeostasis.

very important subcategories are shown on figure 1.3: T helper 1 (Th1) cells mainly stimulate macrophages, while T helper 2 (Th2) cells mainly activate B cells, and the recently identified T helper 17 (Th17) cells release the proinflammatory IL-17 cytokine (Bettelli, Oukka, and Kuchroo 2007). *Cytolytic* T cells (which usually express the CD8 surface molecule) can induce the destruction of their target, in particular virus-infected cells or transplants. *Regulatory* T cells do not stimulate immune effector mechanisms but, on the contrary, downregulate these mechanisms. Regulatory T cells play an important role in immune tolerance and in the prevention of autoimmune diseases. Although the attention has recently been focused on naturally arising $CD4^+CD25^+Foxp3^+$ T cells (Sakaguchi 2005), many other subcategories of regulatory T cells exist (Shevach 2006; see also chapter 3). *Natural killer* T (NKT) cells share characteristics of both innate and adaptive immunity, and they seem to be important in the release of stimulating and regulating cytokines (Bendelac, Savage, and Teyton 2007). Finally, *Gammadelta* ($\gamma\delta$) T cells seem to be crucial for immune mechanisms as diverse as pathogen clearance, inflammation, and tissue homeostasis (Bonneville, O'Brien, and Born 2010). A sixth category (not shown) is *memory* T cells, antigen-specific T cells that persist for long after proliferation in response to a specific antigen. These different subfamilies of T cells will be mentioned regularly in this book.

What happens when a pathogen penetrates the organism? The actors of the "innate" immune system perform the first action: macrophages, granulocytes, the complement, etc. This action occurs most often at the site of the pathogen's penetration. If the pathogen is not eliminated this way, a second action, led by lymphocytes, takes place: antigen-presenting cells (mainly dendritic cells) migrate toward the secondary lymphoid organs, where they present fragments of the pathogen to circulating B and T cells. Some of these cells have very specific receptors for these antigens. Those specific cells multiply and undergo selection mechanisms that lead to expression of receptors capable of interacting strongly with these antigens. These immunocompetent lymphocytes meet up at the site of the infection or lesion, where they organize the adaptive response in association with components of innate immunity: CD8 T cells can directly kill cells infected by a pathogen; CD4 T cells ("CD" stands for *cluster of differentiation*) mediate pathogen recognition by other immune cells; B cells become plasma cells, secretors of antibodies, which interact specifically with the antigen by diverse mechanisms and which promote phagocytosis of the pathogen (a function called "opsonisation").

The Fruit Fly's Immune System

The fruit fly's immune system is rather different from that of the human being, although some elements have been conserved in the course of evolution. The fruit fly can trigger a cellular response and a humoral response. In the *cellular* response, cells called plasmato-cytes phagocytose their target, after a specific interaction with it. The target can be a pathogen or one of the fly's own apoptotic cells. With regard to the *humoral* response, there are two main pathways to immune activation (see figure 1.4). The first, the "Toll" pathway, is mainly directed against fungi, yeasts, and gram-positive bacteria. In this pathway, the Toll transmembrane receptor is activated after binding to a cleaved form of Spätzle (an extracellular cytokine) that is introduced by proteolytic cascades, themselves activated by secreted recognition molecules. The second pathway, called "Imd," is mainly directed against Gram-negative bacteria. It is activated by the direct binding of peptidoglycan recognition proteins ("PGRP") with bacterial activators. These two pathways lead to the activation of different genes and to the synthesis of antimicrobial peptides that are relatively specific to the pathogen (Lemaitre, Reichhart,

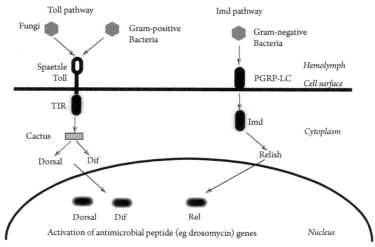

Figure 1.4. The drosophila's humoral immune response: The "Toll" pathway is usually triggered by fungi and gram-positive bacteria. The Toll receptor situated on the cell membrane binds to a cleaved form of Spätzle and then activates a series of events, including the phosphorylation of Cactus, and the release of the transcription factors Dorsal and Dif. These transcriptions factors then move from the cytoplasm to the nucleus and activate antimicrobial peptide genes. The "Imd" pathway is usually triggered by gram-positive bacteria. It is activated by the direct binding of peptidoglycan recognition proteins (PGRP), the most important of which is PGRP-LC, with bacterial elicitors. PGRPs recruit the adaptor Imd. After a complex cascade, the Relish protein is cleaved and the Rel domain translocates to the nucleus, where it activates antimicrobial peptide genes. (After Lemaitre and Hoffmann 2007)

and Hoffmann 1997). This immune system is particularly effective, given the number of pathogens to which the fruit fly is exposed.

The Plant Immune System

The best way to describe the plant immune system in simple terms is to adopt an evolutionary perspective on the interactions between plants and microbes (Chisholm et al. 2006). Plants are constantly exposed to microbes. To exert their pathogenic activity, microbes must penetrate inside the plant, either by the leaves, the roots, openings due to wounds, or natural openings like stomata. Once inside the plant, microbes have to penetrate the cells, getting past the "cell wall," a rigid surface composed of cellulose that wraps around the plant's cells. The moment it attempts to pass the cell wall, the microbe can cause a specific interaction with the host's extracellular surface receptors. As explained above, these extracellular receptors can recognize pathogen-associated molecular patterns (PAMPs). Yet, in the course of evolution, pathogens have developed means of suppressing this immunity directed against PAMPs, by interfering with recognition in the plasmic membrane, or by secreting in the plant cell's cytosol effector proteins that alter its resistance. For their part, plants developed specialized resistance mechanisms to pathogens, which are mainly ensured by NBS-LRR (*Nucleotide binding site-leucine rich repeat*) proteins. NBS-LRR proteins interact in a specific way with pathogenic effectors, or with host proteins that are modified by the pathogen's action (see figure 1.5). The immune response is triggered by the alteration of a nucleotide binding site-leucine rich repeat (NBS-LRR) protein, whether within the direct pathogen detection pathway, or the indirect pathogen detection pathway. This immune mechanism is thus a response to the modification of host's proteins—possibly their phosphorylation (Innes 2011). Moreover, this alteration-based immune response is highly specific. In most cases, when the plant detects the pathogen, it triggers a programmed cell death response that targets the infection site. It can therefore eliminate a large number of pathogens.

The Question of the Criterion of Immunogenicity

The structuring definition of immunology given here is purposely very broad. I have just shown how it may be applied to numerous organisms, from plants to animals. I believe that immunologists could all agree on it. The difficulty, however, is to know whether we can further

Direct pathogen detection

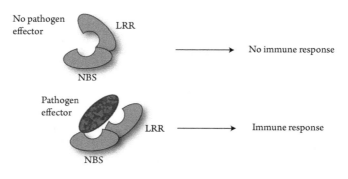

No pathogen effector — LRR, NBS ⟶ No immune response

Pathogen effector — LRR, NBS ⟶ Immune response

Indirect pathogen detection

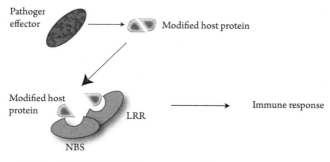

Normal host protein — LRR, NBS ⟶ No immune response

Pathoger effector ⟶ Modified host protein

Modified host protein — LRR, NBS ⟶ Immune response

Figure 1.5. The plant's NBS-LRR immune system: The immune response is triggered by the alteration of a special protein of the plant, belonging to the NBS-LRR (*nucleotide binding site-leucine rich repeat*) family. In the direct pathogen detection, the structure of the NBS-LRR protein is altered directly by the binding of the pathogen effector. In the indirect pathway, the immune response is due to the modification of additional plant proteins. Thus, plants are capable of triggering specific immune responses, always in reaction to the modification of host's proteins. (After DeYoung and Innes 2006)

grapple with this definition by seeking a response to the following question: Immunity is based on specific recognition, but recognition *of what exactly*? The answer usually given by immunologists, at least since Burnet, is that immune components recognize "nonself," which they differentiate from "self." It is critical to understand, then, exactly what these terms "self" and "nonself" mean as they are used in immunology, and what justifies their dominance in the discipline.

WHAT DO THE TERMS "SELF" AND "NONSELF" MEAN IN IMMUNOLOGY?

Immunology, "Science of Self and Nonself"?

Immunology is commonly defined as "the science of self and nonself." Burnet defines it this way in retrospect when he writes: "In 1949, Burnet and Fenner... introduced the concept that the differentiation of self from not-self was the central problem of immunology" (Burnet 1969: 51). In 1949, Burnet and Fenner state, indeed:

> It is an obvious physiological necessity and a fact fully established by experiment that the body's own cells should not provoke antibody formation.... The failure of antibody production against autologous cells demands the postulation of an active ability of the reticulo-endothelial cells to recognize "self" pattern from "not-self" pattern in organic material taken from their substance.
>
> (Burnet and Fenner 1949: 85)

In addition, the idea of self-nonself differentiation is clearly expressed by Burnet in 1962: "The problem of why a chemical pattern is not antigenic in any animal which possesses that pattern as part of its bodily structure is probably the most important of all immunological questions" (Burnet 1962: 36), but also in 1969 in his monumental book on self and nonself, where, relying on skin and tissue transplants, he writes, "The time was ripe to emphasize the importance of 'self' and 'not-self' for immunology and to look for the ways in which recognition of the difference could be mediated." (Burnet 1969: 7). Immunologists in the 1970s and 1980s take up this definition of immunology as science of self and nonself again (Klein 1982). The definition becomes widely accepted, both in the community of immunologists and by the public at large (Wilson 1971; for a recent defense, see Clark 2008). Dausset sees "self-recognition" as a fundamental aspect of immunology (Dausset 1981: 1473) and gives as a definition of immunology "the science of defense against nonself while respecting the self" (Dausset 1990). The example of transplantation is particularly illuminating here: it seems to illustrate perfectly the fact that the organism recognizes its own individuality (its "self") and rejects everything that is foreign to it (its "nonself"). As Burnet writes (1962: 14): "It is as if the body can recognize its own individuality and will accept nothing that is inconsistent with that individuality." Transplant rejection, which constitutes one of immunology's most important themes, seems to make clear why this discipline takes self and nonself as its central objects of study (Kahan 2003). The organism would know its "self" and would accordingly defend it against any threat coming from the "nonself"—that is to say, a *foreign* component, whether this be a microorganism or a donor organ (Clark 2008).

Yet what do the terms "self" and "nonself" mean exactly? It is, at least at first glance, astonishing for a philosopher to notice that a discipline as deeply experimental as immunology uses as its central concepts terms which come from psychology and, before that, philosophy—or metaphysics, to be more precise. Burnet (1940) explicitly admits that the idea to use the terms "self" and "nonself" came to him upon reading *The Science of Life* (Wells, Huxley, and Wells 1929), in which the authors discuss, in a final chapter, the psychological self. In English, the word "self" seems to have first been used in philosophy. According to the *Oxford English Dictionary*, the first occurrence of "self" is found in John Locke's *Essay Concerning Human Understanding* (Locke [1693] 1979). The term "self" used in immunology thus finds its roots in philosophy and, later on, in psychology. Of course, it may be perfectly conceivable that immunologists give a precise experimental definition to this initially nonbiological, nonexperimental notion. The case of transplantation, again, seems to give precise content to the term "self" when it is used in immunology. Nevertheless, a close examination of immunological literature on "self" and "nonself" shows that these two ideas can be understood in strongly different ways that are rarely made explicit. For example, can we say that "self" is a synonym for "organism"? The answer seems to be yes when we look at transplantation, but in other contexts, such as that of immune system development, the organism appears quite distinct from the self, as when immunologists claim that "the organism learns to recognize the self." In other words, should we say that the organism *is* the self, or that the organism *possesses* a self? Furthermore, is the self that which is the organism's "own"? Although in certain languages, especially Spanish, immunological "self" is translated as "own" (*propio*), one's own is more closely associated with "having" (it is what exclusively belongs to me), whereas the self is more closely associated with "being" (it is what defines who I am, my identity). But if there is indeed a difference between one's "self" and one's "own," what significance does this distinction have for a biologist? Another important issue is knowing whether the self is fixed, or "given" (at birth or shortly thereafter), or if it is the product of a progressive construction, possibly by a dialog with the environment. I will return to these questions, in particular the latter, throughout this book. For now, I will give the five main meanings that the word "self" can have in the language of immunologists.

Different Meanings of the Term "Self" in Immunology

There are at least five different meanings of *self* in immunology, though immunologists almost never say explicitly what precise meaning they have in mind. I will return in detail in following chapters to these different meanings, but it is important here to give a first definition in order to establish that the problem of knowing what the immunological notions of *self* and *nonself* refer to has no simple, unequivocal answer:

1. The "self" defined as *the whole organism*. The self would be the sum of the organism's constituents (organs, cells, proteins, etc.), and the immune system would accordingly recognize all entities that differ from these constituents, reacting to the latter and not to the former. In this case, the self is mainly defined at the phenotypic level. The difficulties of this definition are, as I shall show, numerous; they mainly involve the issue of immune tolerance to exogenous entities. If the self is *this* given organism, then how does the self-nonself theory explain the organism's acceptance of a component from a genetically identical individual?

2. The "self" defined as *the individual's genome*. In that case, the self is conceived of at the genetic, not phenotypic, level. This definition accounts, in the field of transplantation, for the compatibility between identical twins. Burnet himself was more and more attracted by the idea of basing the self on genetics. Yet some phenomena of immune tolerance, again, are not conveniently explained if this definition of the self is adopted: How can we understand the absence of rejection with regard to a body that is half genetically foreign, as in the case of maternal tolerance of the fetus? How do we understand, furthermore, tolerance with regard to commensal or

symbiotic bacteria? Again, I will return to these questions in subsequent chapters to demonstrate the fragility of a purely genetic definition of the "self."

3. The "self" defined as *all the markers of the major histocompatibility complex (MHC), called HLA in human beings.* This definition is phenotypic, like the first one, but it is much more precise, since it focuses on tissues. It has often been said, particularly following the work of Jean Dausset on the HLA system, that the MHC constitutes the organism's molecular "identity card" (e.g., Dausset 1981). The "self" would thus be the organism's histological identity. This definition's advantage lies in the very reason that led to its proposal. It allows immunologists to think accurately about transplant rejection and acceptance. It does not, however, get any closer than the other definitions to the understanding of immune tolerance: the fetus's tissues carried by the mother, for instance, are different from those of the organism in which it exists, and yet the fetus is not rejected.

4. The "self" defined as the *group of peptides presented to T cells at the time of their selection,* which occurs in the thymus. I am thinking here of the interpretation of immune "self" proposed by Kourilsky and Claverie (1986), under the name "peptidic self model." T cells recognize the self only in the form of peptides associated with major histocompatibility complex (MHC) molecules, hence the expression of "MHC-restriction." T cells undergo a selection in the thymus, a specialized organ: there, they "learn" to not attack self peptides. Due to the fact that, since Medawar, we know that foreign components introduced in an organism at very early stages of its development can, at least in some cases, be subsequently tolerated by it, the idea of the peptidic self model is that the organism's immune "self" is the group of peptides presented

to its lymphocytes during the selection process. This definition of the self is useful and rich, but it runs into several difficulties, in particular the underestimation of the role of other actors of immunity (notably those of "innate" immunity, such as dendritic cells, macrophages, mast cells, and polymorphonuclear leukocytes) in an immune response.

5. The "self" defined as *that which does not trigger an immune response*. This definition is tautological: it consists of claiming at the same time that the "self" does not trigger an immune response, and that one must call "self" that which does not trigger an immune response (Silverstein and Rose 1997). The result is that all components an organism tolerates are, by definition, an integral part of the "self": the fetus carried by the mother, commensal bacteria, non-rejected parasites, etc. The difficulty then is that the "self" and the "nonself" are reduced to synonyms for "non-immunogenic" and "immunogenic"; conceived that way, the self-nonself theory is purely tautological and ceases to provide an explanation for immunity.

Burnet Faced with the Notions of Self and Nonself

An entire series of questions emerges from these five definitions of the self. Can the meaning of the terms "self" and "nonself" be distinguished, when they are not always specified by immunologists themselves? Can a hierarchy of different meanings be created out of the definitions listed here? Can some be excluded in favor of others? Do any of these meanings lend a solid foundation to the self-nonself theory? Burnet is the one who proposed the self-nonself vocabulary, and then the self-nonself theory, as he was the first to consider these two notions as a *problem*. Therefore, understanding the precise definition of "self" and "nonself," as well as the foundations and subsequent evolutions of the self-nonself theory, requires a conceptual analysis of Burnet's thought. Indeed, it seems crucial to question it in order to grasp what this theory was in the 1950s, and what it has become today. I shall explore the different meanings of the terms "self" and "nonself" while showing how the founder of the self-nonself theory, Burnet, evolved on the question of knowing how to define the self from an immunological point of view. Why does Burnet introduce the concepts of "self" and "nonself" to immunology, and what exact definition does he give to them? How was this still-dominant self-nonself theory elaborated, and what is its empirical basis?

Chapter 2

The Self-Nonself Theory

In the previous chapter, I tried to give a general definition of the terms "self" and "nonself" in immunology. Although these two notions are considered by the majority of immunologists as central to their discipline, their meanings are varied and often imprecise. I will proceed here to a historical analysis of the concepts of "self" and "nonself" in order to clarify their meaning. My main objective is to determine what the terms "self" and "nonself" (or "not-self")[1] meant for the scientist who first introduced these words as part of a scientific *problem*, that is, Burnet. I will begin, however, by showing how the concepts of "self" and "nonself" have a history and how their underlying idea, that what is foreign is immunogenic, had already been expressed by precursors of Burnet, mainly Elie Metchnikoff and Paul Ehrlich. I will therefore also have to establish the nature of Burnet's specific contribution. Not only did he introduce the terms "self" and "not-self," but he also, much more fundamentally, brought about the basic assertion of immune self-knowledge in his efforts to elucidate the mechanisms by which an organism learns to not attack the self. In this he formulated a genuine theory, that of self and nonself. Nevertheless, the distinction between self and nonself was not historically the first problem for Burnet: the inquiry he believed to be fundamental to immunology was that of the establishment of the clonal selection theory of immunity; I will emphasize how this theory, according to Burnet, responds to the problem of knowing how the immune system learns to distinguish self from nonself. The notions of "self" and "nonself" are thus elaborated gradually by Burnet, stemming from the formation of the acquisition of self-recognition as a problem and the proposal of the clonal selection theory. I will show the ways in which the definition Burnet gives of "self" and "not-self" changes between the first significant occurrence (Burnet 1940) and subsequent developments. Only when the definitions of the terms are set, and the theory of clonal selection articulated, is Burnet able to bring a genuine self-nonself theory to the fore. I will examine the foundations of this theory and what experiments came to reinforce it during—and also after—Burnet's active scientific period (roughly from the 1920s to the 1970s). By way of assessment, I will illustrate why the self-nonself theory has continued to dominate immunology from the 1970s to today, while highlighting the important doubts it raised as to the precision of its concepts and its experimental validity.

SELF AND NONSELF BEFORE BURNET

From the end of the nineteenth century to the late 1930s, three broad trends in thought paved the way for Burnet's conceptualization of self and nonself, and strongly influenced the latter: the affirmation of a link between immunity and identity in Metchnikoff's work; the *horror autotoxicus* thesis formulated by Ehrlich; and, finally, experiments on graft rejection and tolerance. The role of these precursors demonstrates that, despite Burnet's repeated statements, he is not the first scientist to have claimed that immunity had a relation to the definition of the organism's identity and individuality.

Immunity, Identity, and Foreign Recognition According to Metchnikoff

By placing phagocytic cells at the heart of immunity, Ilya Ilyich Metchnikoff (1845–1916) gives immunology its first theoretical elaboration. He links immunity and the question of the organism's identity with the concept of *integrity*: every immune reaction is a response to a threat to the organism's integrity, provoking an inflammation (Metchnikoff 1892). These threats can be either exogenous (a foreign body) or endogenous (dying or malignant cells, for example) (Metchnikoff [1901] 1905: 522). Phagocytes are the actors of immunity. These are, according to Metchnikoff, both responsible for the ingestion of foreign entities, as well as the organism's "garbage men," since they also ingest dead cells that are waste from the organism's metabolic activity requiring elimination (Metchnikoff 1884; see also Tauber and Chernyak 1991). The recognition of foreign components is therefore clearly conceived of as a foundation of immunity by Metchnikoff, well before it was by Burnet; it is just that, in Metchnikoff's work, this recognition of what is foreign is but one of the aspects of immunity, the other important one being the ingestion of dying cells by phagocytic cells. It is remarkable that Metchnikoff's argumentation vis-à-vis his contemporaries is not concerned with the homeostatic role of phagocytes of dying cells ingestion, but rather with their possible role in immunity, as if this second aspect were less obvious to the scientific community at that point in time (Metchnikoff [1901] 1905: 420).

The concept of integrity allows Metchnikoff to explicitly articulate the organism's immunity and identity: the latter is defined by the phagocyte, because it is the phagocyte that distinguishes between that which is part of the organism and that which is not and must, as a result, be ingested or destroyed (Tauber 1994: 19). It is by the phagocytes' immune activity that the organism's identity is constructed; that is to say, identity is created by the distinction between what enters into the organism's cohesion and what does not. Metchnikoff is thus the precursor to Burnet in two critical ideas, although these are often credited to Burnet: first, the establishment of a link between an organism's immunity and the definition of its identity; and second, the claim that the penetration of foreign substances into the organism is one of the triggers of an immune response.

Ehrlich and the Horror Autotoxicus as Ancestor of the Self-Nonself Theory

German immunologist Paul Ehrlich (1854–1915) is not content to assert that a foreign substance introduced into an organism triggers an immune response; he wants to know what, in the large majority of cases, makes the organism not mistake its own components as immune targets despite the variety of substances that are capable of triggering an immune response. He responds to this question with his *horror autotoxicus* theory (Ehrlich 1900; Ehrlich and Morgenroth 1901), according to which it is simply inconceivable that an organism could attack itself (*"autos"*) to the point of destruction. The organism must therefore be capable of self-recognition and of distinguishing between what belongs to it and what is foreign to it. As a result, Ehrlich claims that the organism has an identity, and that its immune components are not able to threaten it. The idea of an organism incapable of self-identification in the sense that it might produce "auto-poisons" seemed, in effect, completely "dysteleological" to Ehrlich (1906: 388). The insight behind this claim is that life cannot harm itself. Nevertheless, Ehrlich did not think that autoreactive cells were impossible in principle, though he did affirm that if such cells were to exist, regulatory mechanisms would impose themselves immediately and would suppress the event of pathological autoreactivity (Silverstein 2001). Despite the subtleties of Ehrlich's analysis, the idea attached to his name was indeed that of a strict impossibility of an immune response being triggered against the organism's own components, giving birth to what can rightly be called the "dogma" of *horror autotoxicus* (Moulin 1990). The formulation of this dogma is all the more astonishing given that the existence of auto-antibodies had been demonstrated at the beginning of the twentieth century. Yet this demonstration did not actually destroy Ehrlich's theses, and it did not change the dogma's pertinence in the eyes of immunologists from this era inasmuch as auto-antibodies were seen as pathological exceptions to the organism's normal immune function.

The *horror autotoxicus* thesis strongly influenced not only Burnet, as we shall see, but also all immunologists of the twentieth century, since the idea of a distinction between self and nonself that lies at the heart of the "self-nonself theory" is very often interpreted as the claim of the radical impossibility, except in pathological cases, of autoreactivity.

Transplantation Experiments and Immune Identity

The field of transplantation, particularly after the Second World War, revolutionizes the immunological concept of the organism's identity. Doctors understand that the acceptance and rejection of grafts involves immune mechanisms, in other words, that grafts are an immunological issue; previously, this was not taken for granted. Immunologist Leslie Brent has written a riveting account of the creation of this new field of transplantation medicine (Brent 1997). These discoveries concerning transplantation influenced Burnet considerably as he elaborated the immunological self-nonself theory.

From the end of the nineteenth to the beginning of the twentieth century, the link between transplantation and individuality is confirmed, since the distinction between the results of an autograft (graft of an individual onto itself) and of an allograft (graft of one individual onto another), though assumed much earlier, is at that time clearly established. The mechanisms of this distinction, however, are not well known. From 1913 to 1925, James B. Murphy (1884–1950) performs a series of remarkable experiments (see Murphy 1913) that chiefly concern the allogenic transplantation of tumors. Murphy is the first to clearly confirm that the acceptance and rejection of grafts depends on lymphocytes. Unfortunately, Murphy's experiments attract scant attention from his contemporaries, and it will take many years for it to be "rediscovered" that lymphocytes play a major role in graft rejection.

Leo Loeb's (1869–1959) experiments, and, even more clearly, his theoretical influence, on the idea of a link between transplantation and identity are decisive (Loeb 1930, 1937, 1945). His primary question was that of individual difference between living beings: Loeb highlights the uniqueness of each organism and is thus above all interested in the question of synchronic identity. He shows how immunology in the 1930s, having integrated the already-surprising results of organ transplant experiments, becomes a major biological discipline in order to think about the uniqueness of each living being and degrees of difference between two given organisms, even those within the same species.

> There is inherent in every higher individual organism something which differentiates him from every other individual, which can be discovered by observing the reactions of certain cells and tissues belonging to one individual towards the tissues and cells of another individual of the same species. . . . And not only do these cells recognize the different individuals as such, they do more than that, they recognize, to speak in a metaphorical way, the degree of difference between two individuals as based on their genetic constitution.
>
> (Loeb 1937: 2)

Nevertheless, the person who had the most influence on the elaboration of a concept of immune identity, through the weight of his influence on Burnet's thinking about self and nonself, was Sir Peter Medawar, who was without a doubt the most important figure in the field of transplantation in the twentieth century. Moreover, it is Burnet's interpretation of Medawar's experiments on immune tolerance that earned him the Nobel Prize in 1960, together with Medawar himself.

BURNET'S SELF-NONSELF THEORY

From the perspective of the concept of "self," it is helpful to distinguish two periods in Burnet's thinking: the first, which extends approximately from 1937 to 1945, is that of the affirmation of the distinction between self and nonself; the second, which begins around 1945, is that of the constitution of this distinction as a line of scientific inquiry, wherein lies Burnet's true contribution.

Self and Nonself before the Problem of Self-Tolerance: The "Ecological" View (1937–1945)

Burnet's contribution is not the assertion of the *fact* of distinction between self and nonself, which he always discusses in his work as an obvious experimental given. For example, the difference between self and nonself appears quite clearly in the 1949 work written with Fenner, but it is presented as self-evident, not as a discovery or new claim: "It is an obvious physiological necessity and a fact fully established by experiment that the body's own cells should not provoke antibody formation" (Burnet and Fenner 1949: 85). Until the end of the 1940s, Burnet, if he uses the terms "self" and "not-self," does not take them as the object of his demonstration; they simply serve to designate this undeniable physiological given in which an organism cannot mistake itself as a target for an immune attack.

A close examination of Burnet's writings leads the reader to be wary of Burnet's own retrospective reconstructions. For instance, he writes in his autobiography that he "introduced the concept of the difference between self and not-self" in 1937 (Burnet 1968: 190). In 1969, recalling that his approach in the 1950s consisted of proposing a selective theory of immunity, he writes: "The time was ripe to emphasize the importance of 'self' and 'not-self' for immunology and to look for the ways in which recognition of the difference could be mediated" (Burnet 1969: 7), and, further on: "In 1949 Burnet and Fenner...introduced the concept that the differentiation of self from not-self was the central problem of immunology" (Burnet 1969: 51). Finally, in a remarkable special issue of the journal *Scientific American* dedicated to immunology, and published in 1976 under the direction of Burnet himself, then at the very end of his scientific career, he writes:

> In every discussion of immunity, since Ehrlich first spoke of "horror autotoxicus," the problem of why and how foreign material introduced into an animal provokes antibody production and removal of the foreign material has been balanced by the other problem of why and how the body tolerates its own substance apparently without immune response. For many years I have implied that the basic feature of immunity was the capacity to differentiate between self and not-self.
>
> (Burnet 1976b: 114)

The phenomenon of scientists reconstructing their own conceptualizations and experiments is frequent. It is simply incorrect that Burnet invented the idea of immunological differentiation between the individual and the foreign. And as a matter of fact, in the preceding citation, Burnet is completely conscious of inserting himself into a long line of predecessors, particularly Ehrlich.

Why does Burnet claim to have "introduced the concept of difference between self and not-self" in 1937? He is in fact referring to a text published in 1940, but written in 1937–38 (Burnet 1940), in which he wonders, for the first time, about the mechanisms that allow every organism, no matter how simple, to recognize its own individuality from that which is different from it (Tauber 1994). Noting the ability of an amoeba to digest some external ingredients, particularly microorganisms, without digesting its own substance, he attributes to the amoeba a capacity for differentiation between self and nonself and writes: "The fact that the one is digested, the other not, demands that in some way or other the living substance of the amoeba can distinguish between the chemical structure characteristics of 'self' and any sufficiently different chemical structure which is recognized as 'not-self'" (Burnet 1940: 29).

Burnet's first vision of the self is therefore general to the living world (from unicellular organisms to human beings) and not specific to immunology. The two ideas are complementary: Burnet's idea is that a minimum capacity for differentiation between "self" and "exterior" is indispensable for nutrition, and thus for survival, and that this necessarily has value for all organisms through evolution. It is noteworthy that from the first occurrence of the term "self" it is associated with the term "nonself" ("not-self" in Burnet's wording). If Burnet uses this pair of notions from the start, it is because he wishes to designate a *difference* of physiological behavior: an organism does not react in the same way to its own elements and to those that are outside it.

In addition, it is quite remarkable that Burnet, in his autobiography (Burnet 1968: 23), takes the time to explain the origin of his use of the term "self," which also marks the intellectual context in which he acquires the idea that the organism has an identity. He writes of having found a good portion of his biological inspiration, and even the term "self," in Wells, Huxley, and Wells (1929), as I highlighted in chapter 1. He even acknowledges this intellectual debt in the preface of (Burnet 1940). Wells, Huxley, and Wells propose an ecological vision of the living world, insisting, in particular, on interactions between the organism and its environment. They also suggest viewing the organism itself as an ecosystem, as the place and the product of multiple interactions between different species. Burnet is captivated by this way of seeing, which echoes his own preoccupations at that time with interactions between man and virus. He entitles the first chapter of his 1940 work "The Ecological Point of View" and then introduces the term "self" in the second chapter. Wells, Huxley, and Wells, taking inspiration from Jung, only use this term to refer to the psychological capacity for self-knowledge one finds in human beings and, perhaps, in certain animals. Burnet appropriates the term "self," which he then merges with the ecological vision of the organism, and postulates as a result a capacity in all organisms (the amoeba, for instance) to differentiate between itself and everything that is foreign to it. This "self" is ecologically defined; consequently, it is conceived of as a complex, heterogeneous, and dynamic reality in constant flux. So, the first period in Burnet's conceptualization of the self is above all the assertion of a vision of the organism. In this era, Burnet proclaims a holistic, ecological, "biological" vision, which he pulls from *The Science of Life*. In one sense, Burnet stayed faithful to this last point his entire life: He never departed from the idea that immunological phenomena can only be explained by a global biological concept (Crist and Tauber 2000), which involved him knowing practically all of the biology of his time (something he did to a remarkable degree) and also never forgetting that a biological phenomenon cannot be explained without reference to its evolution.

But little by little, Burnet abandoned the "ecological" vision of the self in order to favor a genetic one. Burnet's "self" during the period stretching approximately from 1937 to the mid-1940s is: (1) organism-based (the self is defined "phenotypically"); (2) ecological (the organism is an "ecosystem"); (3) valid for all organisms, from amoebas to human beings. Subsequently, the self will gradually become: (1) genetically defined; (2) the expression of genetic homogeneity that the immune system must maintain; (3) especially important for higher vertebrates, which alone, according to Burnet, possess true immunity. In the first period, the term "not-self" is present, but it is not a key term; what is important is to point out a vision of the organism that can be understood as one "self." In the second period, the "not-self" becomes the more important term: Burnet wants to show that the immunogenic, that which comes to threaten the organism's integrity, is defined as the foreign in the sense of being genetically different; the "self" is then recognized as that which the organism defends, as opposed to the "not-self," which designates everything that must be rejected in order to survive.

If one sets aside the very terms of "self" and "nonself" that have remained in use until today, Burnet's contribution to a theory of immunity is, in the first period of his activity (until the mid-1940s), negligible: he is content to repeat what his predecessors say, which is that the foreign, and only the foreign, is immunogenic. In reality, Burnet's true contribution is to have elevated differentiation between self and nonself to the level of a scientific *problem*.

Tolerance and the Capacity to Distinguish between Self and Nonself

DIFFERENTIATION BETWEEN SELF AND NONSELF ESTABLISHED AS A PROBLEM

Burnet's contribution consists of making the fact of differentiation between self and nonself move in status from *explanans* to *explanandum*. He is the first to assert that this fact itself needs an explanation: How is it, asks Burnet, that an organism is capable of recognizing self and nonself? And is this "recognition," coupled with a "memory" (since, in certain organisms at least, a second interaction with the same antigen provokes a more rapid and stronger response), innate or acquired?

What ultimately leads Burnet to a true problematization of self and nonself are experiments on grafts and immune tolerance led by Ray D. Owen (1915–2014) and Peter Medawar (1915–1987): these experiments show that the capacity for differentiation of self and nonself is acquired, and therefore that, far from being self-evident, this capacity requires an explanation. In 1944, Medawar shows that an allograft is systematically rejected and that a second graft taken from the same donor is always rejected more quickly and more strongly (Medawar 1944), and deduces from this the fundamental immune mechanism of graft rejection.[2] Meanwhile, Owen (1945) makes a surprising observation on calf fraternal twins: during their entire lives, these calves, which shared the same placenta, keep cells from the other twin and can accept these cells or tolerate a graft of each other's tissues without triggering an immune response. The two calves are called "chimeras"; organisms that possess cells that are genetically different from their own without experiencing immune rejection.

Burnet is aware of these experiments, which he considers extremely important to his own research. From 1949 onward, Burnet and Fenner cite Owen's experiment, though very recent, as a complement to that of Murphy. Burnet will, moreover, make constant reference to it in all his work, as he also will to Medawar's experiments. From this moment, Burnet understands that what had been seen up to this point as a physiological self-evident phenomenon, namely the capacity for an organism to accept its own components while rejecting everything that differs from them, is in reality the result of a complex process of "self-learning" that demands a scientific explanation.

LEARNING SELF-TOLERANCE

Burnet makes a synthesis between observations from embryology (Owen's demonstration of a possible immaturity in the organism's rejection capacity) and immunology (Medawar's demonstration of the role of immune actors in the graft rejection phenomenon). He asks the following question: How does the organism learn to not attack the self, to not trigger an immune response against the self? We have seen that the 1940s and 1950s constitute a key moment in the meeting between immunology and developmental biology: by studying embryos and newborns, researchers understand why differentiation between self and nonself is acquired and not innate. This appears quite clearly in Burnet and Fenner's 1949 work, whose seventh chapter is entitled "Immunological Behaviour of Young Animals."

According to Burnet, Owen's experiments on calf fraternal twins show that an immune tolerance is possible in the embryo but also in the adult; that is, he does not believe that there is anything obvious or immediate in the claim that an organism rejects the foreign while tolerating its own components. In effect, the affirmation of the *horror autotoxicus* principle has already established the general idea: the exogenous is immunogenic, while the endogenous is not. However, since exceptions to this rule are observed in cases of experimental transplantation, Burnet thinks it is necessary to understand how this learning of self-tolerance is achieved in the large majority of cases. As Burnet will highlight in retrospect, the idea that the self must be learned is not at all obvious. Indeed, it seems to be a matter of urgent necessity for every living being that this self-recognition be innate: "At first glance, there seems no reason why the ability of a body to veto immune action against its own components should not depend simply on genetically programmed qualities. It is clearly essential that such tolerance should exist, and it came almost as a surprise that tolerance was not laid down genetically but had to be learned." (Burnet 1976b: 114). Here, Burnet denies the obviousness of *horror autotoxicus* in order to establish it as a scientific question demanding experimentation and conceptualization.

The claim that Burnet makes the move from the fact of distinction between self and nonself to the problem of knowing what makes it possible dovetails with a thesis advanced by Anne-Marie Moulin (1990), according to which Burnet transformed the negative dogma of *horror autotoxicus* into the positive question of how self-tolerance is learned. In effect, Burnet no longer asks "Why is it impossible for an organism to attack itself?" but rather "How does the organism learn to not attack itself?"

As early as 1949, Burnet and Fenner claim that there indeed is self-learning and that immunology must explain the process that leads to this learning. They observe that mammal fetuses and chicken embryos are immunologically immature, since they are not capable of producing antibodies, and write: "the process by which self-pattern becomes recognizable takes place during the embryonic or immediately post-embryonic stages" (Burnet and Fenner 1949: 102).

Burnet and Fenner also conceive of the organism's development as a "hardening" or a "thickening": in the course of development, they assert, "the patterns engraved during embryonic life harden as it were and become permanent possessions" (p. 103). In other words, the motifs present in immature life will harden, thicken, and never be rejected by the organism. This is why, according to Burnet and Fenner, no immune response is triggered against foreign cells (i.e., genetically different cells) that are implanted and established during the embryonic phase. But this also would explain that, if a pathogenic microorganism infects the embryo *in utero*, the animal after birth will be incapable of triggering an antibody response to an injection or infection of this same microorganism. The vision of "self" Burnet proposes is thus dynamic, not rigid as was Ehrlich's: It is no longer a question of simply asserting the impossibility of autoimmunity, but of demonstrating the processes that allow for the acquisition of self-tolerance. That being said, once immune maturity is reached, the definition of the self is concluded, and it can no longer be modified.

The paradox therefore is that Burnet and Fenner approach the problem of self-nonself differentiation by its opposite idea; that is, by the possibility that an immune tolerance exists, that there is an absence of rejection against some genetically foreign components. Burnet and Fenner break down the supposed obviousness of self-nonself differentiation, in order to ultimately better establish its principle: they demonstrate that this distinction is acquired, not innate, and thus that events that occur during the developmental phase of the organism are of major importance for understanding the elaboration of the self. Yet at the same time, they construct immune tolerance as a restrained and marginal phenomenon, because they claim first that tolerance to foreign entities can only be induced during embryonic life, and second, that it is only effected in experimental transplant situations or in the very rare natural cases of chimerism (as with calf fraternal twins). Burnet and Fenner understand, then, that differentiation between self and nonself constitutes a scientific problem, and they strive to propose a hypothesis to explain the acquisition mechanism.

All of this explains why Medawar's and his team's (Billingham, Brent, and Medawar 1953) most famous experiment, four years after the second edition of *The Production of Antibodies*, was seen as a confirmation of Burnet and Fenner's theses. In Burnet's own reconstruction: "It was predicted [by Burnet and Fenner in 1949] that appropriate injections of antigens in the embryo would give rise to subsequent tolerance of that antigen. Eventually the prediction was abundantly fulfilled" (Burnet 1969: 51; see also Brent 2001). In effect, Billingham, Brent, and Medawar show that it is possible to introduce an active tolerance to foreign tissues in mice, as long as the foreign cells are introduced to the host early, during fetal life, i.e., before immune maturity—experiments that have then been repeated and confirmed (e.g., Steinman, Hawiger, and

Nussenzweig 2003; Dakic et al. 2004). They describe this phenomenon as "the exact inverse" of "actively acquired immunity" and as
a result propose the term "actively acquired tolerance." Medawar,
who during his scientific career had occasion to underscore the
contribution of transplant experiments on the definition of individuality (Medawar 1957), also showed that an extension of "self" was
possible in experimental conditions, exactly as Burnet and Fenner
had suggested a few years earlier (Medawar 1960).

Burnet's Real Theoretical Fight: The Clonal Selection Theory

While the historiography of immunology traditionally presents
Burnet as the inventor of self and nonself, what is striking while
reading the articles and books he published from 1955 on is that
his true theoretical engagement seems to lie elsewhere. Burnet is
certainly trying to convince his readers of one theory's validity, but
it is not the self-nonself theory. In all his texts from 1957 to 1970,
Burnet strives to promote his "clonal selection theory." As he writes
in his autobiography: "Rightly or wrongly I regard the development
of the clonal selection theory of immunity as my most important
scientific achievement." (Burnet 1968: 190).

One objection could crop up here: Doesn't the close examination of Burnet's self-nonself concepts as central to his theoretical
elaboration of immunity risk misrepresenting his thought? In reality, as we shall see, this objection can be dismissed, because clonal
selection theory is a response to the problem of knowing how an
organism learns to recognize the "self," as Burnet himself affirmed.

I shall begin by asking what comprises clonal selection theory.
In *The Production of Antibodies* (Burnet 1941), Burnet tries to determine how the synthesis of antibodies is achieved. It is precisely to
this question that the clonal selection theory endeavors to respond.

Up until the mid-1950s, immunology is dominated by the "instructionist" theory, whose most important representative is Linus Pauling
(Pauling 1940). According to this theory, when an antigen penetrates
an organism, it creates in the antibody a specific complementary
structure; the antibody adopts a spatial configuration molded on the
"template" of antigenic determinants, which allows this antibody to
react specifically with the antigen: the result is an immune response
targeted at the antigen's destruction (Silverstein 1985: 271). Before
this theory took hold, immunology was dominated by the thesis
offered by Ehrlich in 1897, which may be considered the first selective theory of antibody formation (Silverstein 1999). Nevertheless,
Breinl and Haurowitz (1930) opposed Ehrlich's theory on the basis
that it did not allow consideration of the apparently infinite number
of antibodies. Pauling's instructive theory responded precisely to
this issue. Burnet, however, had always been skeptical of instructionist theories: in the late 1920s, he expressed strong doubts about the
idea that a bacterium's resistance to a bacteriophage would be the
result of an "instructive" process, and he proposed instead a selective explanation (Burnet 1929). As for antibodies, Burnet will oppose
the "instructionist" theory with a "clonal" theory: antibodies exist
before the antigen enters the organism, and cells carrying antibodies with the highest specificity and affinity with the antigen are simply selected over other less competent immune cells. Burnet, despite
his own "selective" intuitions as early as the 1920s and 1930s, always
claimed a debt to Niels Jerne's (1911–1994) groundbreaking article
(Jerne 1955) in developing the principle of clonal selection theory. In
1957 Burnet publishes a text with a significant title: "A modification
of Jerne's theory of antibody production using the concept of clonal
selection" (Burnet 1957). Simultaneously, David W. Talmage puts
out a review article (Talmage 1957), in which he suggests, without
completely explaining, a thesis close to Burnet's, although each man
initially ignored the other's argument.

Burnet is intrigued by the clonal explanation of immunity advanced by Jerne, but it seems insufficient to him on several experimental points. Its major shortcoming for Burnet is that it views selection at the level of the antibody. For Burnet, it is clear that it is immune *cells* that are selected based on the antibodies they carry on their surface, and not the antibodies themselves. Burnet's theory is that it is necessary to propose *"the existence of multiple clones of globulin-producing cells"* (Burnet 1959: 54, author's own emphasis; contemporary immunologists would say "antibody-producing cells"). Immunocompetent cells would carry molecules on their surface similar to antibodies that are synthesized and react with the antigen, allowing its destruction. These cells would then be selected on the basis of their receptors' specificity with regard to antigenic determinants. There is no process of "instruction" there, says Burnet, but only a process of Darwinian selection at the cellular level (Burnet 1959: 64; see also Silverstein 2003).

Burnet publishes his text, well aware of its rather unorthodox nature, in a little-known scientific journal, *The Australian Journal of Science.* He considers the risk that his peers may view his hypothesis as scientifically aberrant, and he believes it better to publish in a minor journal. Burnet clearly describes his logic in his autobiography (Burnet 1968: 206): if his proposition were proven true, he would be considered its inventor; if it were quickly invalidated, few researchers in England or the United States would know about it.

At that time, the instructionist theory was, however, well established. In 1957 Burnet begins the main theoretical fight of his life, which is to convince his peers that only a selective theory of immunity could hold up. This battle remains quite vivid in Burnet's reflections in the 1970s (Burnet 1970, 1976b). Burnet says, no doubt in a somewhat exaggerated manner, that over nearly ten years (approximately from 1957 to 1967) he experienced many difficult moments trying to making himself understood. In retrospect, in a text written to celebrate ten years of his clonal selection theory, Burnet even compared himself to Galileo confronting the Church (Burnet 1967: 3).

The theory of clonal selection of lymphocytes has, without a doubt, been the main theoretical struggle in Burnet's scientific career. Moreover, it is one that ends in a clear victory for him: from the end of the 1960s, the large majority of immunologists rallied to this theory. It would be erroneous, however, to believe that the problem of clonal selection theory excluded the problem of defining the self because the former would be more fundamental. The two concerns are intrinsically linked. Clonal selection theory takes as its task the explanation of immune response to an antigen, as well as immune tolerance and self-learning, i.e., the fact that initially (during the fetal or immediate postnatal period) an organism tolerates any substance (endogenous or exogenous), but then rapidly acquires the competence to only accept "self" and to reject "nonself." This brings him to a fundamental point, which Burnet asserts in his 1959 work: to the idea of competent immune cells' selection following the penetration of an antigen into the adult organism, one must add another idea, that of a selection during the embryonic stage that eliminates immune cells that recognize self components. Under one single theory of clonal selection, Burnet associates a hypothesis related to development with another related to immunity. With regard to development, Burnet's idea in 1959 is that receptors borne by immune cells are randomly generated, which makes the appearance of autoreactive cells possible, but that these autoreactive cells will be eliminated during the maturation process: "Self-not-self recognition means simply that all those clones which would recognize (that is, produce antibody against) a self component have been eliminated in embryonic life. All the rest are retained" (Burnet 1959: 59). Thus, Burnet himself repeatedly states that clonal selection theory is, from his perspective, the only immune theory capable of explaining self-tolerance. This is very clear in Burnet's 1969 work, entitled, as we know, *Self and Not-Self*, in which he asserts: "There is more than one explanation of tolerance but the existence of the phenomenon was the stimulus that led to the conception of the clonal selection approach to immunity" (Burnet 1969: 25).

Clonal selection theory serves therefore to justify both the selective mechanism of immune response, as well as self-tolerance. As such, it will be at the core of Burnet's response to the problem of recognition between self and nonself: "It was immediately evident that such an approach [the clonal selection theory] provided an alternative mechanism by which the amount and type of antibody and immunocytes could be adjusted to the current needs of the body. It also provided the simplest possible interpretation of how the body's own constituents are shielded from immunological attack" (Burnet 1969: 12).

Evolution of the Concept of Self in Burnet's Thought: From the Ecological Self to the Immunogenetic Self

We have seen that Burnet borrowed the notion of "self" from *The Science of Life* authors, who proposed an ecological conception of the organism. Despite Burnet's interest in this ecological vision, he progressively breaks with it in favor of a genetic concept of the self. From this perspective, it is important to note that Burnet was particularly well informed about genetic advances of his time, and that he followed with close attention all scientific developments of the 1950s and 1960s concerning the structure of DNA and its expression in protein synthesis. In his writings, Burnet often cites key actors in these discoveries (Watson and Crick, later Monod and Jacob, etc.), upon whom he never hesitated to comment and whose results he sometimes even remarkably anticipated. Burnet is among the first to see the potential in the development of an "immunogenetic" branch of immunology. In the late 1950s, Burnet thinks that an organism's identity is essentially defined by its genome. Of course, immune identity also plays a crucial role for the organism in the rejection of pathogens, self-tolerance, and prevention of autoimmunity; however, immune identity is, in the vast majority of cases, only the reflection of genetic identity, as the organism learns to recognize its own components that are, with only very rare exceptions, genetically defined. The individual genome is thus placed at the heart of the organism's identity, as well as at the core of the criterion of immunogenicity.

Although Burnet took as a starting point Medawar's experiments on tolerance, which showed that the immune "self" cannot be reduced to the genetically defined "self," he considered these processes of immune tolerance as exceptions resulting from simple experimental manipulations to the general principle of immunity, which is clearly genetically grounded. Burnet's writings contain an evolution toward the idea that the organism's individuality is found in its genes and that an effective immune system is one that is capable of learning to recognize protein products of this genetic individuality. Burnet, beginning in 1959, conceives of the immune "self" as the reflection of each organism's genetic individuality, except in experimental manipulations (Burnet 1959: 33–35). Immunogenicity depends on the absence or presence of an antigen during the immune system's maturation: if, during this period, a foreign antigen is present in the organism, it will be tolerated for the organism's entire life. Now, in almost all cases only those components created by its own genetic material are present in the organism during the immature immune period, so it is a perfectly legitimate approximation to consider the immune self as the antigenic reflection of genetic individuality. These "self" antigens will not trigger an immune response. Anchoring the immune self in genes also explains why grafts between identical twins are tolerated.

This is confirmed in all later texts, beginning with Burnet's lecture on December 12, 1960, after receiving the Nobel Prize and which is, significantly, the first of his texts to include the term "self" in its title (Burnet 1960). In 1962, Burnet reexamines results in the field of experimental transplantation in the first half of the twentieth century through the lens of this genetic anchoring: "It is one of the concise statements of modern immunology that the body will accept as itself only what is genetically indistinguishable from the part replaced" (Burnet 1962: 13–14). In 1969, Burnet clearly explains genetic anchoring and its possible exceptions. For him, transplantation experiments show what the self is and on what basis the self must be understood: The heart of the mechanism is *the identification of a difference that is genetic in origin*: "In the first place we have the demonstration that for a tissue to be rejected it must be recognizably *different* and that the differences involved are genetic in origin" (Burnet 1969: 24, author's italics). Yet Burnet adds that there can be graft tolerance between two individuals who are genetically *different*. This is possible in the case (and only this case) where exchanges of individual components have taken place during the organism's development, *before* complete immunocompetence is established:

> Genetic identity is not, however, a *necessary* condition for skin graft acceptance. The intermingling of placental blood of two dissimilar twins is a natural experiment which shows that tolerance of another individual's tissues is possible if the body has experienced the presence of foreign cells from a period early in embryonic life. From this deduction the whole topic of *immunological tolerance* has developed and in a sense the present hope that organ transplantation will one day be regularly possible.
>
> (Burnet 1969: 24)

What is clear here is that immune tolerance is an important object of study for Burnet, but also that he does not imagine that it can be produced *after* the period of "immaturity," that is, the fetal

or the neonatal period. For him, tolerance concerns only development, in the narrow sense of the period that leads the organism to
adult age.

This movement from an ecological "self" to a genetic "self"
renders the term more scientifically precise. However, it simultaneously leads to a "hardening" of the term, since, as I shall show,
Burnet limits the possibility of immune tolerance to the brief window of immunity acquisition that corresponds to the fetal or neonatal period. Burnet goes from an open self to a closed one, from the
concept of an organism as the unity of a plurality interacting with its
environment and susceptible to its influences, to viewing the organism as a homogeneous unit produced by its genes and whose identity
is defined in a premature developmental period, then whose integrity must be defended against any environmental threat. This precise definition of the self that stems from genetic anchoring allows
Burnet to put forth the first genuine *criterion of immunogenicity*, that
is, the first true explanatory immunological theory.

Self-Nonself Differentiation as a Criterion of Immunogenicity

THE FIRST ELABORATION OF A CRITERION OF IMMUNOGENICITY

Burnet's ambition is strong. He is looking to determine a *criterion of
immunogenicity*, that is, a response to the question of knowing under
what conditions an effective immune response occurs. His objective
is not to offer a mere description, it is to find an *explanatory* theory
of immunity. The idea is to be able to anticipate what entity triggers an immune response if it is introduced into an organism and to
explain why. Burnet thus proposes the first scientific theory offering
a criterion of immunogenicity. Burnet's answer is that an organism's
immune system is able to differentiate between self and nonself.

The two central assertions of his theory are as follows: (1) the organism triggers an immune response against any foreign ("nonself") entity; (2) the organism does not trigger an immune response against any entity that belongs to it ("self").

This immunological theory, truly explanatory and predictive, allows for the explanation both of a phenomenon perceived at that time as obviously immunological—the defense against pathogens—as well as of a phenomenon that had up to this point not been seen as such, the rejection of grafts. The key in triggering a response of immune rejection would be, in all cases, the foreign, i.e., that which is different from the organism's individuality.

WHAT ARE THE EXPERIMENTAL FOUNDATIONS OF BURNET'S SELF-NONSELF THEORY?

The question that arises here is not to know what the experimental foundations of self and nonself theory are *today*, but rather those on which Burnet was able to rely in order to assert his theory in 1949, as well as to maintain and reinforce it until the end of his active scientific period in the early 1970s.

The main experimental foundations of Burnet's theory involve grafts. In the large majority of cases, in vertebrates, especially in humans, a graft between two allogenic individuals is rejected. Now, whereas one can interpret the immune response to pathogens in many ways, immune rejection of transplanted organs or tissues that could be useful to the host seems to have as its only explanation the capacity of the organism to recognize its individuality and to combat everything different from it. Burnet's reliance on transplantation to anchor his theses also involves, without being paradoxical, tolerance, that is, the acceptance by an organism of foreign organs or tissues. As has been shown, Burnet accompanied Owen and Medawar's discoveries, sometimes even amazingly predicting them. Each time, it was, in Burnet's view, a matter of exceptions to the rule of self and nonself, exceptions that are almost always artificial as a result of experimental manipulation. But what is important for him is that they help us to understand the process of self-nonself differentiation.

Furthermore, beginning with the end of the 1950s, Burnet advances the idea that immune cells play an essential role in immunity, specifically suggesting that antibodies are not the only actors of immunity. He theorizes that at least certain immune cells undergo their maturation in the thymus. In the early 1960s, several experiments, notably that of Miller on thymectomy (Miller 1962), confirm yet again Burnet's theory.

Burnet thus experimentally grounds his hypothesis in two fields that, in his time, and partly thanks to him, are converging: that of pathogenic resistance and that of graft rejection. He submits that the second field should be considered as strictly relevant to immunology, and he proposes his self-nonself theory as a structuring explanation for all available experimental data. All experiments preceding Burnet's theoretical elaboration and all those that follow him seem to go in the same direction, namely toward the need to posit a capability for any organism to distinguish between what is itself and what is not, all immune response being based on this discrimination.

As we have seen, the principle that it is the foreign that is immunogenic predates Burnet's own contribution. Therefore, from the moment that the only truly controversial proposition—that is, Burnet's clonal selection theory—becomes accepted, a double "freezing" occurs: the theory of clonal selection and the principle of self-nonself differentiation "are no longer a theory but a fact" (Klein [1982] 1990: 335). Burnet himself confirms in 1967 that his impression is that for ten years immunology has followed the path he had traced—before pulling back in modesty and saying that immunology would have practically been the same without his influence (Burnet 1967: 4). As for Jerne, after the same 1967 conference, he offered the significantly titled text "Waiting for the End," whose overarching idea was that immunology found its theoretical stability around Burnet's clonal selection theory and it was then only a matter of waiting for young immunologists to discover the remaining molecular details:

> Sir Macfarlane Burnet must have been pleased not only to witness at this symposium the vindication of his Clonal Selection Theory of Acquired Immunity, but also to see how his stimulating ideas have led to a great proliferation of immunologists, and to know that the fate of immunology is deposited in so many capable hands. As this younger generation of professionals is pressing rapidly toward the definitive solution of the antibody problem, we older amateurs had perhaps better sit back, waiting for the End.
>
> (Jerne 1967: 601)

At this point, self-nonself discrimination appears as quite self-evident, as an idea that no longer requires examination.

TO WHICH ORGANISMS DOES THE
SELF-NONSELF THEORY APPLY?

In 1940, Burnet claims that the distinction between self and nonself must exist from the level of the amoeba up, since all forms of digestion rest on such a mechanism. One problem arises, however: Does self-learning truly exist in all organisms, even the simplest ones? Must we say that any organism is capable of self-nonself differentiation? That all organisms are endowed with immunity? Burnet did not ignore the fact that in plants and certain animals, grafts "took" much easier than in vertebrates. To take one example in which Burnet himself was interested, a colony of *Botryllus schlosseri* can fuse with a colony that is not strictly identical from a genetic perspective (Burnet 1971).

The problem of knowing which organisms have an immune system is even more important for Burnet because evolution is central to his thinking. In this, his intellectual approach breaks with the great majority of immunologists of his time (and also, to be honest, of our time), who are little inclined to link their work to evolutionary issues. Burnet is always concerned with not leaving immunology in a strictly physiological framework and with situating it in the context of evolutionary biology, something which, according to him, all life sciences should do. He is unwavering in his conviction that immunology only takes on its true, full meaning with regard to evolution (e.g., Burnet 1962: 2). Now, it is an equally fundamental issue for the perspective I defend in this book to delineate the domain of life to which an immunological theory claims to apply.

In reality though, Burnet vacillated between these questions without ever really deciding between them. His evident wish is to consider immunity of all organisms, from unicellulars to mammals. Nonetheless, Burnet focuses so clearly, at this point in his scientific career, on antibody production (for which he offers his master theory of clonal selection), that he often admits that his immunological theory (e.g., Burnet 1969) only applies to vertebrates, or even just mammals. Indeed, the transition from the ecological interpretation of self and nonself to the interpretation based on clonal selection theory also marks the transition from a general biological theory for all organisms to a theory centered on organisms that have antibodies, that is, higher vertebrates. From this point of view, the 1960 Nobel conference is, again, a revealing text: along with the genetic turn as far as the definition of the self is concerned, it also offers a phylogenetic turn concerning the capacity of self-nonself differentiation, by clearly asserting that immunology is concerned with vertebrates only (Burnet 1960: 689). And yet, the end of the text is dedicated to the evolutionary origins of immunity. What can explain this?

The first explanation is that all his life Burnet remained fundamentally a medical doctor, and as a result, immunity was for him first and foremost, if not exclusively, a science that took human beings as its goal, even if it sometimes involved using animals in order to understand human immunity. This "medical" view of immunology is still largely shared today. Burnet focuses on so-called "adaptive" immune phenomena, which is to say phenomena related to the possession of immune "memory," and suggests that one can only ever truly talk about immunity when such adaptive immunity exists, thus only in jawed vertebrates.

The second explanation is that even if he places the self at the core of immunity, Burnet dissociates the ability to differentiate self and nonself, on the one hand, and the capacity to trigger an immune response, on the other. In 1960, Burnet states that, from an evolutionary perspective, protection against pathogens cannot be at the origin of a self-recognition mechanism, whereas the inverse is probable. Thus, for Burnet, a "self" and "other" distinction mechanism exists in all living beings and must have appeared very early in evolution. Burnet (1971) illustrates this idea with self-pollination inhibition seen in certain plants: he suggests that such phenomena show that self-nonself recognition is ubiquitous in the living world, contrary to immune defenses that only truly exist in jawed vertebrates. Burnet's position does not hold up to scrutiny today, for we know immunity exists in practically all organisms and goes back to very ancient evolutionary times (see chapter 1). However, Burnet, after a brief period of uncertainty, offers a position which is coherent: having defined immunity as the possession of antibodies, he limits it to a very small number of living beings, higher vertebrates; but, as a good evolutionist, he tries to determine what evolutionarily useful capacity gave birth to immune recognition, and he claims that this capacity is self-nonself differentiation.

CONTEMPORARY IMMUNOLOGY AND SELF-NONSELF THEORY: BETWEEN HEGEMONY AND DOUBT

The Dominant Theory from the 1960s to Today

During Burnet's active scientific period, which lasted until the late 1960s, several experiments came to confirm his hypotheses and insights. It is even more astonishing that, from the end of the 1960s until the end of the 1980s, many experimental discoveries appear to follow Burnet's ideas. This is particularly the case with the discovery of the major histocompatibility complex in humans and the clarification of the mechanisms that generate diversity in antibodies.

HISTOCOMPATIBILITY: THE IMPORTANCE OF DIVERSITY AND POLYMORPHISM

The discovery of the major histocompatibility complex (MHC) is an extremely important step in the consolidation of the self-nonself theory. George D. Snell (1903–1996) first discovered MHC in mice, where it was called "H-2" (Snell 1948). The term "histocompatibility antigens" is the basis for a convergence between genetics and immunology. MHC is then discovered in human beings by Jean Dausset, which earned him the Nobel Prize in 1980 with Baruj Benacerraf (1920–2011) and George D. Snell (see Dausset 1981). Thanks to MHC molecules, the organism's cells continuously display on their surface peptidic fragments of the proteins they synthesize. For instance, a cell infected by a virus will show abnormal peptides on its surface, which, after having reacted specifically with T cells, will trigger an immune response. MHC is thus rapidly understood as the organism's "identity card," and T cells as its surveillance system. Immune cells would have as their task the constant verification that the organism's cells are expressing "self peptides" on their surface, and to destroy them if this is not the case. MHC plays a decisive role in graft rejection, of course: the question of "histocompatibility" is that of the possibility of finding tissues belonging to one individual but that are acceptable to another individual. Yet work on MHC demonstrates the difficulty of compatibility: apart from identical twins, it is extremely rare to find two "compatible" individuals for transplantation. Of course, there are degrees of compatibility, and certain grafts are commonly practiced today, which shows that some transplantations are quite possible. However, they almost always require the use of immunosuppressive drugs and often end up being rejected. The discovery of MHC shows the polymorphism of living beings, their genetic and phenotypic uniqueness, and comes then to reinforce the self-nonself theory. It is when he describes the function of MHC that Dausset (1990) writes that immunology is "the science of the defense against nonself, while respecting the self."

THE IMPORTANCE OF T CELLS AND THE
DEMONSTRATION OF THEIR CLONAL ELIMINATION

Closely related to the work on MHC, one of the biggest breakthroughs in immunology in the 1970s and 1980s is the assertion of T cells' central role in immunity, an idea that is still maintained today. This is a rather recent notion: before it, immunologists were first and foremost concerned with *antibodies*. Now, early work on grafts, and then the work on MHC (which is closely linked to the problem of grafts) shed light on the cellular aspect of immunity, no longer in the sense of phagocytes (as had been the case when, after Metchnikoff, one would speak of "cellular" immunity), but in the sense of T cells.

During the 1970s and 1980s, a series of experiments (see specifically Zinkernagel et al. 1978 and Kappler, Roehm, and Marrack 1987) resolved the question of knowing how "self-tolerance" is possible in a way that obviously went in the direction of Burnet: the demonstration of the elimination of T cells that carried "self" antigens on their surface.

THE GENERATION AND DIVERSITY OF ANTIBODIES

The other major discovery that comes to strengthen the self-nonself theory is the response that Susumu Tonegawa (born in 1939) brought in the mid-1970s to the question of antibody diversity (see Tonegawa 1974, 1983), which earned him the 1987 Nobel Prize. The problem had been raised with Ehrlich's selective theory: How are antibodies, necessarily limited in number, able to recognize an infinite number of antigens? Echoing this difficulty, another doubt had appeared as more and more threatening to immunology in the 1970s: How is it possible to assume that the organism's immune cells carry receptors capable of recognizing any antigen—the number of possible antigens in nature being practically infinite—whereas the number of an organism's genes is necessarily finite? Little by little it appeared that an organism's genes were not numerous enough to code for such a diversity of immune receptors. Many biologists had observed that antibodies had different sequences, and thus corresponded without a doubt to different genes, but the mechanism of this generation of new genes was not clearly identified. In discovering the mechanisms of the generation of diversity, Tonegawa made credible the idea of a random generation of immune receptors; he showed that a small number of genes was sufficient to explain the synthesis of an immense immune repertoire. The main mechanism Tonegawa demonstrated is that of somatic recombination: the variable region of each antibody is in fact coded by many gene fragments (not just by one), which are then reassembled by somatic recombination.

If one takes into account all diversity mechanisms at the level of antibodies, one arrives at a theoretical number in human beings of 5×10^{13}.

Seen as validations of Burnet's concepts, these discoveries strongly contributed to immunologists' adoption of the self-nonself theory, which from this point on, was rarely questioned again.

THE NUMEROUS CONCEPTUAL CONTINUATIONS
OF THE SELF-NONSELF THEORY

In 1986, Philippe Kourilsky and Jean-Michel Claverie proposed the "peptidic self" model, which constituted an extension of Burnet's vision, all while offering an important evolution in the definition of the self (Kourilsky and Claverie 1986). The immune "self" must, according to them, be defined in relation to T cells, which are the main cells guarding the organism and which play the role of conductor in any adaptive immune response. Since T cells recognize the self only in the form of peptides associated with MHC molecules, the immune self must not be, in their view, defined genetically, but rather "peptidically": the self is the set of peptides present in the thymus at the moment of immune maturity, which, as a result, will not be the target of an immune response in the organism.

Numerous other conceptualizations show the vigor of the self vocabulary and of the principle of self-nonself differentiation, for instance Klas Kärre's theory of "missing self" applied to "natural killer" cells (Kärre 1985), or the idea put forth by Janeway that the immune system would react to the "infectious nonself" but would tolerate the "noninfectious self" (Janeway 1989, 1992). These different terms and hypotheses attempt to elaborate on Burnet's theory, and sometimes to modify it, but they fall clearly in his wake.

Thus from the 1960s to the 1990s, the immunological concepts of self and nonself are solidified, in the twofold sense that they become accepted by the entire community and that they are no longer questioned, but accepted as the adequate explanation of immunogenicity.

THE DOMINATION OF THE SELF-NONSELF THEORY IN CONTEMPORARY IMMUNOLOGY

The self-nonself theory continues to dominate today's immunology: setting aside a small number of exceptions, to which I will return, contemporary immunological articles all affirm its relevance. This is quite evident in numerous experimental articles and reviews (e.g. Guillet et al. 1987; Nossal 1989, 1991; Anderson et al. 2002; Wraith 2006; Jiang and Chess 2009), and even clearer when immunologists take the time to present their conception of immunity. In one issue of *Seminars in Immunology* dedicated to the immunological self and nonself, in 2000, the editors write: "Given that everyone agrees with the proposition that a biodestructive defense mechanism must make some kind of self-nonself discrimination, there will hopefully be little disagreement that this requires a specificity element that recognizes chunks of the biological universe with sufficient precision to distinguish those parts that belong to self from those that belong to nonself" (Langman and Cohn 2000). This confident remark, though fortunately criticized inside the same volume (Cohen 2000; Grossman and Paul 2000), is frequent among today's immunologists (e.g., Langman 1989; Cohn 1998a, 1998b; Clark 2008; Cohn 2010).

Doubts and Theoretical Disinclination

FIRST DOUBTS REGARDING SELF AND NONSELF

The doubts, or at least the first modifications, regarding the self-nonself theory came from Burnet himself; unfortunately, these revisions were paid scant attention by his contemporaries who remained, in their great majority, dedicated to the strict differentiation between self and nonself. For example, toward the end of his life, Burnet endeavored to understand the immune response as a check on abnormal endogenous modifications: mutations due to age, cancer, etc. (Burnet 1970, 1976a, 1978). He thus outlined a new role for autoimmunity: immune cells could, in certain circumstances at least, react to antigens generated by the genetic "self."

This reevaluation of autoimmunity followed along "systemic" positions, which is to say views that insisted on the idea of an immune "system," and which all more or less referred back to Jerne's founding article (Jerne 1974). According to these positions, autoreactivity is a normal process in organisms, since the immune system must react with the organism's endogenous components in order to be able to ensure that nothing abnormal occurs. Autoreactivity, which only becomes autoimmunity strictly speaking in the case of an autoimmune disease ("self"-destruction), would thus no longer be considered impossible. In the 1990s, extending in part Jerne's ideas, the American immunologist Polly Matzinger proposed her "danger model," whose main idea is that the immune system reacts not to "nonself" but to any "danger," endogenous or exogenous, in the organism (see chapter 5).

MAINTAINING THE "SELF" WHILE FAVORING MOLECULAR DESCRIPTIONS

Faced with the expression of such doubts or such concurrent hypotheses, today's immunologists remain in most cases attached to the self-nonself theory, but give it a loose, purposely imprecise meaning, considering that the true challenge lies in the elucidation of molecular mechanisms of immunity. In other words, the theoretical framework used would matter little, as long as it would allow for advancement in molecular immunology.

One of the most significant aspects of this position is the development of an analysis in terms of activating and inhibiting signals, which is extremely widespread today. Within such an analysis, it is no longer a matter of posing a radical alternative between the situation where one immune cell is activated by nonself and that where there is no activation: on the contrary, it would be necessary to think in terms of *degrees* of activation, as each immune actor integrates, on the model of neural cells, multiple activating and inhibiting signals, its final state resulting from the synthesis of these integrations. The functioning of NK cells, cellular interactions, the growing taking into account of the role of cytokines all enter into this view. The vocabulary and principles of the self-nonself theory are thus maintained, but only in relation to this interplay of activating and inhibiting signals. This view of immunity is certainly very accurate: all immune response is complex and contextual; it depends on numerous signals coming from the cellular and molecular environment. I have nothing to object to this description, except that it is precisely only a description, and not an explanation, of the immune response—which runs counter to the ambition of Burnet's self-nonself theory.

THE RETURN TO THE UNQUESTIONED
SELF-NONSELF DISTINCTION

To put it rather bluntly, in today's immunology, self-nonself differentiation, and more generally the question of a criterion of immunogenicity, have ceased to be problems, and became again what they were before Burnet, which is to say allegedly self-evident facts requiring no further examination. This is why contemporary immunologists no longer conceptualize self and nonself, or even occasionally no longer "believe" in it, but with very rare exceptions (to which I will return in the next chapters), do not offer any alternative concepts and continue in their publications to think in these rather imprecise terms. This makes the caricature of Burnet that contemporary immunologists often draw all the more harmful, whether it is done by proponents of the self-nonself theory or by its critics. It is my hope that the close reading of Burnet's texts offered here proves that he was an exceptional scientist and more generally an exceptional thinker, at once inventive and cautious, a specialist (in virology and immunology) and a generalist, steadfast in his conceptual contributions (the clonal selection theory) and open to new proposals (tolerance, innate immunity). Burnet's thinking is even more rich and subtle than a number of today's immunologists would believe, and even than their own views of immunity, often little-conceptualized and little-theorized, with few exceptions. For my part, it is precisely because I think that Burnet was correct in his most fundamental ideas, starting with the determination that a criterion of immunogenicity is possible and the conviction that immunity must be understood from an evolutionary perspective, that I wish to show that his answers to these problems can no longer be considered valid today. Accepting and taking seriously the immunological problems Burnet raised, let us try to show the ways in which his proposed solutions are inadequate, and then with what it would be useful to replace them.

Chapter 3

Critique of the
Self-Nonself Theory

In the previous chapter, I showed the conceptual and experimental foundations of the self-nonself theory, a theory that still dominates today's immunology. In this chapter, I detail the reasons why I think this theory is inadequate. First, I will present experimental data that demonstrates a large number of exceptions to the idea of self-nonself differentiation. My critique will proceed along two main lines, autoreactivity (i.e., an immune reaction to the "self") and tolerance (i.e., the absence of an immune reaction to the "nonself"). Then, I will show that the self-nonself theory is connected to a reductive vision of immunity in which only higher vertebrates possess an immune system. Finally, I will analyze the conceptual vagueness of the self-nonself theory and will suggest that it is necessary to propose another theory, which I will try to do in chapters 4 and 5.

NORMAL AUTOREACTIVITY AND
AUTOIMMUNITY

Contrary to the assertions of the self-nonself theory, the immune system reacts continuously to endogenous antigens (i.e., to the "self"). This normal autoimmunity appears as one of the main components of the organism's homeostasis, defined here as a group of internal regulatory mechanisms following a disturbance. The idea of normal autoimmunity was suggested in the 1970s, in the framework of the "systemic" vision of immunity put forth by Niels Jerne. The experimental arguments this vision relied on, however, were very thin. Here, I take up this idea of normal autoimmunity, but with a fresh perspective, in order to show how recent experiments come to establish it.

An Important Distinction: Autoreactivity and Autoimmunity

Does autoimmunity have to be thought of as a malfunctioning, an exception to normal immune function? This is what the self-nonself theory proposes. Yet we shall see that this is erroneous. It is indeed necessary to distinguish three concepts that the self-nonself theory does not differentiate: *autoreactivity, autoimmunity,* and *autoimmune disease.* To truly understand the difference among these three notions, let us begin by distinguishing, in general, the immune *reaction* from the immune *response.*

IMMUNE REACTION AND IMMUNE RESPONSE

An immune *reaction* designates the biochemical interaction between an immune receptor and its ligand. In contrast, an immune cell triggers an immune *response* if and only if it is *activated.* Immune activation is the stimulation of the cell, possibly followed by differentiation and proliferation. Immune-activating mechanisms can be put into two broad categories:

1. *Effector* mechanisms lead, directly or indirectly, to the destruction of the target. This destruction can be achieved by phagocytosis (ensured by macrophages, dendritic cells,

and neutrophils, among others), lysis (by CD8 T cells or natural killer (NK) cells, in particular), stimulation of other cells (e.g., activated "helper" CD4 T cells can stimulate macrophages and B cells), or recruitment of molecules that ensure or favor the lysis of the target or its phagocytosis (molecules of the complement, for instance).

2. *Regulatory* mechanisms, by contrast, lead to the *inhibition* of immune effector mechanisms. These regulatory mechanisms include regulatory T cells, the HLA-G molecule, numerous cytokines like TGF-β, etc.

An immune response is preceded by an immune reaction, but not every immune reaction leads to an immune response. In most cases, an indispensable condition for releasing an immune response is a strong specificity, affinity, and/or avidity binding between the immune receptor and the ligand involved. In addition, it is important to distinguish the triggering of an immune response at the level of *one* cell and at the *systemic* level: an immune response occurs when an immune cell triggers activating mechanisms (whether effector or regulatory mechanisms), but one speaks of a systemic immune response if and only if several immune components are activated and take part in a collective immune response.

CONSEQUENCE: THE DIFFERENCE BETWEEN
AUTOREACTIVITY, AUTOIMMUNITY,
AND AUTOIMMUNE DISEASE

Following the distinction between immune reaction and response that I have just put forth, I now turn to *autoreactivity*. There is *autoreactivity* in all cases where an immune cell's receptors interact with an endogenous antigen. In contrast, the term *autoimmunity* refers to the situations where immune activation (destruction of the target or inhibition of this destruction) is triggered against an endogenous antigen. Autoimmunity thus presupposes autoreactivity, but the inverse is not true. Finally, one speaks of an *autoimmune disease* when an organism's immune system triggers a destructive response against its own organs or tissues. The development of an autoimmune disease presupposes autoreactivity and autoimmunity, but there again, the inverse is, quite fortunately, false. Indeed, autoreactivity, autoimmunity, and autoimmune disease are not three increasing levels of one same interaction, but rather three processes that are distinguished by their result: autoreactivity occurs when there is interaction with endogenous components, autoimmunity when there is destruction of endogenous components or inhibition of this destruction (in other words, activation of effector or regulatory mechanisms as defined above), and autoimmune disease when there is destruction of the organism's tissues or organs.

What is Normal Autoreactivity?

THE CONSTANT SURVEILLANCE OF THE ORGANISM'S COMPONENTS

At the basis of normal autoreactivity lies the fact that the immune system performs a constant surveillance over all the components that make up the organism. Immune cells effectively interact with the other cells in the organism and, if they are activated, they can provoke the destruction of the target.

The immune surveillance system comprises many complexes, the best studied of which is the major histocompatibility complex (MHC). MHC includes the "HLA" (human leukocyte antigens) system in humans. I will first concentrate here on the human immune surveillance system before moving on to the examination of other species. In humans, most of the organism's cells constantly present at their surface, thanks to what are called HLA class I molecules, peptides originating from the proteins that they synthesize. Thus, for example, a cell infected by a virus presents viral peptides at its surface, and consequently can be eliminated by the immune system. T cell receptors interact with complexes made up of an HLA molecule and a peptide. If the interaction is characterized by a strong specificity, affinity, and/or avidity, then effector mechanisms may be triggered.

The immune surveillance system rests on several other mechanisms. One is the response due to NK cells. These cells—which play a decisive role in graft rejection, the protection against cancers, and also against pathogens—respond to any cell that does not express HLA class I molecules. In other words, according to the suggestion of Klas Kärre, they respond not to the presence of "nonself," but to the absence of a typical "self" molecule (Kärre 1985).

Note that immune antigen-presenting cells (APCs), for their part, display at once on their surface both HLA class I and class II molecules. The latter allow the so-called "professional" presentation of antigens, i.e., the possibility for immune cells that carry them, to activate other immune cells, and, in some cases, to initiate an effector immune response that involves numerous constituents.

In humans, therefore, the immune system, which is made up of many cellular subpopulations (antigen-presenting cells, T cells, NK cells, etc.) and of molecules (cytokines, etc.) constantly provides surveillance of all elements present in the organism.

What about this surveillance system in other species? As we shall see, immune surveillance mechanisms exist in all plant and animal species, although not all have a major histocompatibility complex. Normal autoreactivity based on the existence of an MHC exists, however, in many vertebrates and invertebrates. The H-2 system of the mouse, for instance, is practically identical to the HLA system. Another example is that of the protochordate colonial organism *Botryllus schlosseri* and its FuHC (*Fusion/Histocompatibility*) system (De Tomaso et al. 2005), to which I will return later on.

LYMPHOCYTIC SELECTION: A WINDOW OF REACTIVITY

As we have just seen, the immune system permanently watches over the organism. The T cell receptors interact with HLA molecules presenting the organism's endogenous peptides, which is to say that there are constant immune reactions to "self" components. As a general rule, they are not reactions of strong intensity, or else they might lead to the destruction of the organism's normal components. Yet how is it that T cell receptors never (or only very rarely) trigger a strong response against such normal components? This is the old question, raised by Ehrlich's *horror autotoxicus* and certain aspects of Burnet's theory, of knowing why, even though immune receptors are produced randomly, autoimmune destructions are not more frequent. The response to this question is, as Burnet perceived, that T cells in the thymus undergo a selection (in the sense of an elimination).

For a long time after Burnet, immunologists thought that any immune receptor likely to react with a "self" antigen was necessarily eliminated in the thymus, hence an immune repertoire untainted by any autoreactive receptor. Yet two discoveries in the mid-1990s have shown that this idea was not correct. The first was the demonstration of the relative imperfection of the mechanism by which significantly autoreactive lymphocytes are eliminated: many lymphocytes that react quite strongly to "self" antigens survive the process of elimination in the thymus (Bouneaud, Kourilsky, and Bousso 2000). Admittedly, taken by itself this discovery simply underscores that no biological mechanism is perfect, which is a well-known fact. However, it led immunologists to seriously envision the necessity of peripheral regulatory mechanisms. This consideration has particularly contributed to the emergence of research on regulatory T cells, to which I shall return later on, since they also challenge the self-nonself theory.

The second discovery was that a T cell in the thymus is selected if and only if it reacts weakly—and not if it does not react at all—to the self antigens presented to it. Indeed, in the thymus, some endogenous antigens are presented to maturing T cells by the cortical epithelial cells, which are thymic professional antigen-presenting cells. More precisely, endogenous antigenic peptides are presented by MHC molecules to T cell receptors in the course of maturation (a T cell does not interact directly with a peptide, but with a complex made up of an MHC molecule and a peptide). If the T cell does not react at all with the MHC + endogenous peptide complex, it dies by apoptosis; if it reacts very strongly, the same is true. In the case of B cells, whose receptors bind directly to the antigens that they meet (without an association with an MHC molecule), a similarly weak autoreactivity is necessary for their maturation, which takes place in the bone marrow (Melamed et al. 1998). As a result, during its maturation, a lymphocyte survives only if it reacts weakly to endogenous components presented to it, hence the notion of a "window of reactivity": in order to be selected in the thymus (in the case of T cells) or in the bone marrow (in the case of B cells), a lymphocyte must carry on its surface receptors that bind to endogenous antigens that are presented to it with a weak specificity, affinity, and/or avidity (Ashton-Rickardt et al. 1994). Consequently, a certain dose of autoreactivity is not only possible in a healthy organism, but also truly indispensable to its survival.

AUTOREACTIVITY AT THE PERIPHERY

What is known as "central" immunity refers to immune phenomena produced in central lymphoid organs, namely the bone barrow and the thymus, the organs in which lymphocyte maturation takes place. On the other hand, "peripheral" immunity takes place in the whole organism, but is mainly triggered in the peripheral lymphoid organs (lymph nodes, spleen, tonsils, and lymphoid tissues associated with mucus membranes, like Peyer's patches in the gut), and implies mature immune cells.

In the course of the past fifteen years, to the surprise created by the demonstration that weak autoreactivity is required in the thymus was added an even larger surprise concerning autoreactivity at the periphery. An immune cell that is not regularly stimulated by "self" antigens in the peripheral lymphoid organs dies by apoptosis (Freitas and Rocha 1999). We now know that certain cellular interactions are as fundamental to the maintenance of T cells in peripheral lymphoid organs as they are to their selection in the thymus. In other words, the same type of interactions produced in the thymus at the moment of T cells selection occurs at the periphery between T cell receptors and complexes made of an MHC molecule and a peptide. Far from being fundamental only to the initial selection of T cells, these interactions remain essential throughout the entire life of T cells, as these interactions alone guarantee their survival.

One can deduce from this that continuous (i.e., regularly repeated) interactions of intermediate strength between T cell receptors and MHC + peptide complexes are necessary to T cells' survival. These interactions produced at the periphery are, like all those that take place in the thymus, interactions between T cells and endogenous components ("self" components): there is nothing exogenous in the individual's MHC molecules and in its own peptides that are presented at the surface of the organism's cells. Only the continuous interaction of T cells with "self" components can ensure their survival. This truly shows how erroneous the "self-nonself" dichotomy is, for not only are immune system cells *able* to react to the "self," but they also will not survive *unless* they continuously react to the "self." T cells' autoreactivity is thus a necessary condition for the continued existence of the immune system.

Do forms of autoreactivity similar to that found in T cells exist for other immune actors? The answer is clearly yes, including for actors of so-called innate immunity. This is the case with Toll-like receptors (TLR) carried by antigen-presenting cells, especially dendritic cells. These receptors not only ensure the recognition of microbial pathogenic patterns, they also intervene in the surveillance process of the organism's components; in particular, they react to patterns expressed by the organism's dying or damaged cells. The idea that TLRs recognize only the "infectious nonself" and never the "self" is thus increasingly thrown into doubt: it seems, on the contrary, that reactivity to endogenous entities is normal for TLRs (Beg 2002; Marshak-Rothstein 2006).

Natural killer (NK) cells also exercise a permanent surveillance of antigens carried on the organism's cell surfaces. As I have highlighted above, NK cells respond to any cell that does not express MHC class I molecules. They therefore participate in the normal autoreactivity described here (Raulet, Vance, and McMahon 2001).

Another important example of cells that react with "self" and not with "nonself" is that of natural killer T (NKT) cells (CD1d-restricted T cells). These cells are not activated by the recognition of foreign products, but by the conjunction of two signals, both provided by a dendritic cell: first, an endogenous antigen, and second interleukin 12 (IL-12), synthesized by the dendritic cell activated by a pathogen's presence. Consequently, NKT cells, which are essential immune cells in responses to pathogens (notably viruses), tumors, etc., do not interact with just any "nonself," but receive endogenous signals coupled with signals resulting from an antigen-presenting cell having interacted with a pathogen (Bendelac, Savage, and Teyton 2007).

Autoreactivity of innate immune components is by no means limited to higher vertebrates. For instance, normal autoreactivity exists in the fruit fly (Lemaitre and Hoffmann 2007): the two main paths to immune activation, the *Toll* path and the *Imd* path, perform surveillance over antigens present in the organism and can release activating immune responses against certain endogenous antigens, in particular at a wound site, but also against a tumor or even against certain endogenous DNA.

Thus, the surveillance function of the immune system implies a constant autoreactivity. This is true for innate, as well as adaptive, immunity components and across phyla.

At this stage, an objection is possible: the autoreactivity I am discussing seems to involve only *interactions* between immune cells and endogenous antigens. In other words, there would effectively be autoreactivity (i.e., a simple *reaction* with self components), but not autoimmunity (i.e., the triggering of an immune *activation* against these components). In reality, though, a whole series of activation mechanisms are at work on endogenous components of the organism.

What Is Normal Autoimmunity?

"Normal autoimmunity" refers to the immune-activating mechanisms that target the components of a healthy organism. As we shall see, these autoimmune responses occur in every organism; they are even necessary to its survival because they ensure vital homeostatic functions. These are true immune *responses* and not mere *reactions*, as previously defined: in the case of normal autoimmunity, endogenous antigens trigger an immune activation. I analyze here the diverse manifestations of this normal autoimmunity, in particular the functioning of phagocytic cells and regulatory T cells. In each case, I ask whether the mechanisms I describe exist widely in the living world or are limited to a small number of species.

PHAGOCYTIC CELLS, THE ORGANISM'S "GARBAGE MEN"
Phagocytic cells, particularly macrophages, carry on their surface receptors that react with endogenous antigens (Savill et al. 2002; Taylor et al. 2005; Jeannin, Jaillon, and Delneste 2008). These cells that can respond to pathogens also play the role of

"garbage men," meaning that they get rid of the organism's waste, for instance dead cells, as Metchnikoff had already observed. One million cells die every second in our bodies (Green 2009). These cells are "engulfed" by phagocytic cells, typically macrophages. A macrophage phagocytoses a dead cell just as it would phagocytose a "foreign" entity as a bacterium, for instance (though the consequences of these engulfments may, of course, be very different, in particular because apoptosis is usually not proinflammatory). Thus, macrophages, and phagocytic cells in general, ensure an effector immune response to components that can in no way be defined as "nonself."

In which species can this kind of autoimmunity due to phagocytic cells be found? It is now clear that it exists in many species. In the fruit fly, for example, specialized cells called "plasmatocytes" carry out phagocytosis (Lemaitre and Hoffmann 2007). These cells eliminate bacteria, as well as apoptotic cells. A particular process exists in plants that could appear similar to the ingestion of dead cells that has just been described. In reality, it is a slightly different process, by which the plant triggers a strong apoptotic mechanism against its own cells found around the zone of a pathogenic infection (Taiz and Zeiger 2006). This is referred to as "hypersensitive response." By killing certain cells of its own, the plant deprives the pathogen of nutrients and prevents its spread. In other words, the plant triggers a severe autoimmune response that destroys a part of itself to ensure the survival of the rest of its components.

REGULATORY T CELLS

The growing understanding of the role of regulatory T cells reflects the realization that it is equally important for an organism to regulate and end the immune response as it is to trigger it. An uncontrolled immune response would have devastating consequences, especially due to its proinflammatory and ultimately destructive effects. Many immunologists had, in the 1970s, discussed "suppressive cells," but the idea had subsided before coming up again in the course of the last ten years.

Regulatory T cells (T_{Reg}) inhibit the activity of other immune actors. As mentioned in chapter 1, many populations of regulatory cells exist,[1] but I will focus here on those that are the best known and seemingly play the most important immunoregulatory role: lymphocytes carrying the two molecules CD4 and CD25, and expressing the transcription factor Foxp3 (*forkhead box protein 3*) (Sakaguchi 2005). These lymphocytes are thus called $CD4^+CD25^+Foxp3^+$ cells. They constitute 5% to 10% of the total of mature T cells. They perform an inhibiting immune activity directed against other immune cells. Many regulatory T cells are "self" cells that react with other "self" cells, leading to the regulation of an immune response. More precisely, the most commonly held hypothesis today is that two subpopulations of $CD4^+CD25^+Foxp3^+$ regulatory T cells exist, "natural" regulatory T cells and "induced" (or "adaptive") regulatory T cells (Bluestone and Abbas 2003). The first are selected in the thymus on the basis of their capacity to respond strongly to "self" antigens, and their activation depends on the exposure of these same antigens at the periphery (von Boehmer 2003). The "induced" regulatory T cells are generated at the periphery, either from natural regulatory cells or from what are normally effector lymphocytes. Generation at the periphery depends on the context of antigen presentation, which may be a tissue antigen or a foreign antigen.

The great majority of regulatory T cells, then, do not follow the principle that immune cells always respond to "nonself": on the contrary, they respond to "self" components in the form of an inhibition (i.e., negative regulation).

Regulatory T cells offer one of the most convincing arguments against the self-nonself theory's proposition that the immune system does not respond, except in pathological cases, to the organism's endogenous constituents. Shimon Sakaguchi, the contemporary immunologist who has worked the most on understanding the role of regulatory T cells, readily acknowledges that the self-nonself theory has prevented immunologists from seeing what has been right in front of them all along. Sakaguchi insists that when one studies the immune system, it is imperative, together with the "self-knowledge" mechanism (the Greek *gnothi seauton*), to also take into account regulatory mechanisms—but this time, instead, according to the Greek adage "nothing in excess" (*meden agan*) (Sakaguchi 2006).

OTHER AUTOIMMUNE MECHANISMS
It is increasingly clear that, in addition to $CD4^+CD25^+Foxp3^+$ cells, other cells respond to "self" components, as was recently shown to be the case for Th17 cells. Th17 cells are interleukin 17–producing $CD4^+$ helper T cells, which play a critical role in the early phase of the innate immune response by recruiting neutrophils and inducing the production of antimicrobial proteins and inflammatory factors from resident cells (see chapter 1). These cells were proved to be selected in the thymus on the basis of their *strong* affinity for self-peptides. Moreover, they can respond to endogenous antigens at the periphery and fine-tune the overall immune response (Marks et al. 2009). The set of self-responding cells also include CD1d-restricted NKT cells, $CD8\alpha\alpha$ intraepithelial lymphocytes, and $\gamma\delta$ T cells.

Even if T_{Reg} and other cells just discussed are only present in higher vertebrates, immune inhibition exists in all life. In the fruit fly, for example, several immunoregulatory mechanisms exist, such as the "Cactus" inhibitor molecule or certain PGRP amidases (Zaidman-Remy et al. 2006; Bischoff et al. 2006). In plants, several regulatory mechanisms intervene, especially in the case of the LRR (*leucine-rich repeats*) domains of NBS-LRR proteins (*nucleotide-binding site–leucine-rich repeats*) (see chapter 1). The LRR domain negatively regulates certain immune signals, as demonstrated by the fact that one deletion of this domain provokes the constitutive activation of defense responses (DeYoung and Innes 2006). It is extremely important to note that these mechanisms are preserved in the course of evolution, from plants to animals.

In conclusion, immune regulatory mechanisms make it clear that some immune cells respond to "self" components, and that they exist in many different forms in the plant and animal kingdoms. In reality, the ubiquity of such regulatory mechanisms is not surprising given the damage that an immune response can inflict on an organism.

Result: It Is False That the Immune System Does Not Respond to "Self"

Contemporary data thus shows that autoreactivity is not only possible but necessary in central and peripheral organs alike, and that the organism triggers numerous activating responses against endogenous components. Furthermore, I have demonstrated that such normal autoreactivity and autoimmunity mechanisms very likely exist in all multicellular organisms (plants, invertebrates, vertebrates). The first claim of the self-nonself theory, which states that the organism does not trigger an immune response or a reaction against "self" components, is thus false. Reactions and responses against "self" are actually necessary to the organism's proper functioning.

To be sure, I am not denying here that most organisms do not self-destruct and, in particular, that they do not develop severe autoimmune diseases. But I do not think that the idea of self-nonself differentiation can explain the difference between normal autoimmunity and autoimmune disease since, in both cases, an immune activation is triggered by the organism's endogenous components. Of course, I am not suggesting that there is no difference between normal autoimmunity and autoimmune diseases. Autoimmune diseases presuppose autoreactivity and autoimmunity, but go further, since they constitute a dysfunction with regard to normal autoreactivity and autoimmunity. In presenting my continuity theory as opposed to the self-nonself theory, I will show how I conceive of autoimmune diseases. What is important here is to underscore how the *horror autotoxicus* dogma formulated by Ehrlich prevented immunologists who, following Burnet, adopted the self-nonself theory, from working on conceptual distinctions, however essential they were. The relative scarcity and the clearly pathological nature of autoimmune diseases have led these immunologists to reject, for a long time, the possibility of normal autoreactivity and normal autoimmunity as necessary to the organism's proper functioning. In particular, the idea that the "self" would be incapable of triggering an immune activation was, undoubtedly, a major sticking point to understanding the role of regulatory T cells whose importance I have just emphasized.

IMMUNE TOLERANCE

In its broadest sense, immune tolerance designates the absence of effector (i.e., destructive) immune response to an antigen. According to the self-nonself theory, the organism triggers an immune response of rejection against any foreign (nonself) entity. In these conditions, immune tolerance to exogenous entities, which was broadly conceived of by Burnet and his successors as the absence of rejection of a foreign component, could only be seen as an exception, quite rare and limited, to the rule of self-nonself discrimination. In this context, the term "immune tolerance" primarily referred to the period of immune immaturity found in many animals, during which foreign elements may be accepted by an organism because it is not yet immunocompetent (Burnet and Fenner 1949; Medawar 1960). As a result, given that such a presence of foreign components in the fetus or the newborn was seen as extremely rare in nature, the expression "immune tolerance" served, for proponents of the self-nonself theory, to mean "self-tolerance," that is, the organism's acquisition of the ability to recognize its own components without triggering a destructive immune response against them (the learning of the "self-recognition") (Burnet 1969: 219).

Yet, immune tolerance is in fact quite frequent in nature and far from limited to one initial period of immaturity. Surprisingly, immune tolerance understood as a state of tolerance to exogenous entities is not at the core of any overall conceptualization by immunologists or even of any philosophical reflection focused on immunology (in particular those of Alfred Tauber and Anne-Marie Moulin). Most of today's immunologists continue to consider tolerance as an exceedingly rare and abnormal phenomenon in nature. This relative oversight reflects in part the fact that studies on immune tolerance are a relatively recent trend; most of the experiments discussed here date from the last ten or fifteen years. Still, immunologists' attachment to the self-nonself theory has also contributed to the lack of attention given to immune tolerance. I am pursuing here two goals: on one hand, to give a precise definition of immune tolerance that does not confuse it with immunosuppression (i.e., clinical immunosuppressor mechanisms); on the other hand, from this definition, which relies on numerous converging experimental data coming from immunology, microbiology, and developmental biology, to offer a philosophical analysis of immune tolerance showing the frequency of this phenomenon and the necessity of taking it fully into account in order to evaluate the relevance of the self-nonself theory.

What Is "Immune Tolerance"?

The term "immune tolerance" was initially broadly defined as the absence of rejection, in a given organism, of an entity which is nonetheless foreign to it, that is to say, genetically different from it (Burnet 1969). This definition created a problem, however, since it led to the regrouping under the same term "tolerance" of both (1) non-rejection due to an absence of appropriate immune cells (either because they have not yet achieved their development, as in fetal or immediate postnatal tolerance, or because of the use of immunosuppressive drugs), as well as (2) immunoregulation, or the inhibition of destructive immune responses due to the activation and to the differentiation of certain immune cells. It is essential to distinguish between immunosuppression (definition 1) and tolerance (as defined in definition 2) by showing that tolerance is fundamentally a process of active regulation.

To really understand this distinction, it is useful to take as a starting point the experimental domain that, thanks to Owen, Burnet, and Medawar, lies at the core of the immune tolerance issue, namely that of transplantation (Medawar 1960). In the case of the human being, Burnet's thesis on self and nonself may seem for the most part correct, because the possibility for allogenic grafts that would be perfectly tolerated without immunosuppression is, for the moment, an idea rather than a reality. In human beings, autografts (graft of an individual onto himself) and homogenic grafts (between identical twins) are tolerated, but allogenic grafts (between two genetically different individuals) are almost always rejected. Of course "compatibility" plays an important role: individuals differ more or less from the point of view of their histocompatibility system (HLA system), and doctors work hard to find the best donor able to allow a successful graft to the donor patient. However, even when there is MHC compatibility, minor histocompatibility antigens can provoke a rejection. Remarkable progress having been made in the course of the past fifty years on the possibility of preventing graft rejection with immunosuppressive drugs, one might be tempted to consider this phenomenon as a case of immune tolerance. But in reality, immunosuppressive drugs destroy immune receptors that would respond to donor antigens. Thus, it is not tolerance in the strict sense of the term, but simply a deletion of immunocompetent cells. Tolerance is the product of a process of immunoregulation, that is, the result of an equilibrium of activating and inhibiting mechanisms that eventually leads to the acceptance of a foreign entity.

Likewise, recent progress has been made in the field of chimerism induction, which has been considered by some authors as a case of immune tolerance (Fehr and Sykes 2004). Clinically practiced chimerism rests on a technique that consists of transplanting hematopoietic stem cells in the recipient before performing an organ or tissue graft. Three forms of clinically induced chimerism exist (Claas 2004): "macrochimerism" (total irradiation of the recipient in order to kill its immune cells and transplantation of the donor's bone marrow); "mixed chimerism" (irradiation of the organism and transplantation of a mixture of its own bone marrow with that of the donor, which results in the organism's own newly appeared T cells being selected in the thymus on the basis of both its own antigens and donor antigens, the recipient thus becoming specifically tolerant to any donor tissue graft [Sykes 2001]); and finally "microchimerism" (the introduction of a very small amount of the donor's hematopoietic cells before performing the organ or tissue graft, which, in certain cases, creates a specific tolerance to donor antigens).

These experiments on clinically induced chimerism show that, contrary to what Burnet and practically all immunologists thought until very recently—that immune maturation was a unique event that occurred once and for all in the fetal or immediate postnatal period—the adult immune system *can* be modified (Waldmann 2002). This possibility of redefining an individual's immune identity even in adult age supports the increasingly expressed idea that development lasts throughout the individual's lifetime, even if the adult's immune plasticity is not as important as that of the fetus (or, depending on the species, the newborn). Nevertheless, these experiments on the redefinition of the individual's immune receptors must not be considered as examples of immune tolerance, since immune tolerance is, as I have said, a positive process of cellular activation and differentiation that leads to the inhibition of a destructive immune response.

I have thus removed from my definition of immune tolerance two phenomena (immunosuppression and chimerism induction) that could have wrongly appeared as forms of this tolerance. With my more restricted definition of the term "tolerance" in mind, I will now demonstrate my thesis that immune tolerance is a frequent biological phenomenon. This argument does not at all mean that anything can be immunologically accepted or tolerated by a given organism, but it demonstrates rather that to understand the process of acceptance and rejection by the yardstick of the concepts of self and nonself is not satisfactory. After having examined recent evolution in the field of transplantation that tends to show the possibility of an immune tolerance, I will analyze two phenomena that make it obvious that an organism does not reject every foreign entity: fetomaternal tolerance and the tolerance to microorganisms and macroorganisms.

Graft Tolerance: The Field of Transplantation

The difficulty of achieving allogenic grafts apparently reinforces the self-nonself theory: it seems that each individual is unique and, when it comes to transplantation, can only tolerate its own "self" (or components from an identical twin), unless it receives immunosuppressive treatment. And yet three experimental facts challenge this claim: the existence of immunoprivileged organs, the role of regulatory mechanisms in graft tolerance, and finally the fact that, in other species than humans, graft rejection does not happen in the same way.

THE CASE OF IMMUNOPRIVILEGED ORGANS

Immunoprivileged organs are those that, when transplanted from one individual to another, allogenic, individual, do not trigger an immune response of rejection or trigger a very weak one. This term "immune privilege" was first proposed in 1953 by Rupert E. Billingham, Medawar's student, with regard to corneal grafts (Simpson 2006). Numerous sites have since been called "immunoprivileged," since it appeared that they did not release an immune response even in the case of the graft of a "foreign" organ or tissue: the cornea, the brain (or even, according to certain researchers, the entire nervous system), the testicle, the fetus tolerated by the mother, the pouch situated in the hamster's cheek, etc. It is important to note that, in many cases, notably that of endocrine tissues, immunologists finally realized that immune responses in fact did take place, but that they were simply slower and weaker than in other sites of the organism. However, the cornea, for example, is still considered today to be a paradigmatic example of an immunoprivileged organ.

In order to explain the phenomenon of immune privilege, the first proposed hypothesis was that the immune system could not access these particular sites in the organism whether because they would not have been vascularized or because they would have been protected by barriers that immune cells could not cross. The hypothesis was that there was no immune response simply because the immune system did not have access to these tissues. Now, in the past fifteen years, in particular with the research on regulatory T cells and the HLA-G molecule, immunologists have moved from the concept of immune privilege as a passive phenomenon to the idea that it is instead an active phenomenon, implying multiple regulatory mechanisms, in particular T_{Reg} cells, and cytokines like TGF-β and Fas/Fas ligand (Mellor and Munn 2006).

Consequently, the term "immune privilege" seems today to converge with that of "immune tolerance" while still remaining a separate concept: immune privilege designates certain states of tolerance observed in transplanted tissues or organs. But instead of continuing to explain this tolerance by the immune isolation of the sites involved, today we know that the mechanisms that allow immune privilege are certainly very close to those that ensure, for instance, tolerance of commensal microorganisms (Mellor and Munn 2006).

THE ROLE OF REGULATORY T CELLS AND HLA-G IN GRAFT TOLERANCE

Beginning in the 1990s, researchers demonstrated the important presence of regulatory T cells in the site where a graft took place (Qin et al. 1993; Graca, Cobbold, and Waldmann 2002). These cells were mostly $CD4^+CD25^+Foxp3^+$ cells. The role of regulatory T cells in graft tolerance is clearly established in mice (Wood and Sakaguchi 2003). For instance, it was shown that the suppression of regulatory T cells in mice increased graft rejection, and, inversely, that regulatory T cells introduced with naive T cells in a syngeneic mouse having undergone an allograft increased the graft's survival. As for human beings, most specialists think that $CD4^+CD25^+$ regulatory T cells carry out the same function (Wood and Sakaguchi 2003; Waldmann et al. 2006), but this has only been demonstrated in a small number of cases, which suggests more complex activation mechanisms. Whatever the case may be, these observations on regulatory T cells have contributed to the modification of the old "dogmas" of transplantation immunity (Claas 2005).

Several characteristics of regulatory T cells relative to grafts have been brought to light (Waldmann et al. 2004; Waldmann et al. 2006): first, their receptors are specific to donor antigens presented by the host's antigen-presenting cells; second, they are dependent on continuous exposure to the antigen in order to remain active; third, they ensure, by a phenomenon called "linked suppression," the extension of the suppressor mechanism to other antigens expressed within the tolerated tissue; fourth, they extend the dominant tolerance to other naive T lymphocyte cohorts, a process called "infectious tolerance" (Qin et al. 1993). This last phenomenon constitutes a remarkable case of immune tolerance: some "foreign" antigens can be tolerated by an organism thanks to an active process of stimulating its regulatory T cells, the latter being able to extend their tolerance capability, starting with one initial antigen, to numerous other antigens, imposing a universal tolerance with regard to a grafted tissue or organ. Taken together, the four mechanisms by which T cells ensure tolerance to exogenous antigens show that immunogenicity depends on the integration of activating and inhibitory signals, and not on the origin (endogenous vs. exogenous) of the antigen.

The involvement of regulatory T cells in many immune tolerance phenomena has even led certain researchers to view this tolerance due to regulatory T cells as a case of "immune privilege." It seems that, in a manner similar to that of so-called "immunoprivileged" organs, the graft itself actively participates in its acceptance by the immune system by stimulating, notably via cytokine synthesis, the host's regulatory T cells (Cobbold et al. 2006).

Researchers have of course speculated about therapeutic applications of this observation that regulatory T cells are present in grafts. The aim would be to stimulate the regulatory T cells of the organism undergoing the transplant in order for these cells to inhibit the destructive immune response. Waldmann et al. (2004: 124) use the term "negative vaccination" to designate the selective induction or expansion of regulatory T cells in autoimmune diseases, allergies, transplantation, and other forms of immune pathology. The assumption of this therapeutic research is that not only can the organism's repertoire of reactivity be partially redefined but also that the type and the intensity of effector immune responses triggered by an organism can be modified by inhibiting this activity with the help of regulatory cells. Nevertheless, it is important to underscore that, in human beings, researchers do not consider using only regulatory T cell stimulation, without immunosuppressive drugs. It is thus a promising future technique that could reduce the side effects of immunosuppression (or even eliminate these effects altogether), but which still has to prove its worth. It is reasonable to believe that researchers will make major theoretical and therapeutic discoveries in the field of the immunoregulation of immune response to transplants by regulatory T cells in the not-too-distant future.

In addition to regulatory T cells, another immunoregulatory mechanism, that of the HLA-G molecule, could play a major role in graft tolerance. What exactly is the HLA-G molecule? The HLA system molecules comprise two classes: class I molecules are expressed on the surface of most of the organism's cells, while class II molecules are only expressed on the surface of certain cells that play a role in immunity. The HLA-G protein is a slightly polymorphic HLA class I molecule. Unlike classic HLA molecules, its function is not to present peptides. According to Carosella and colleagues, HLA-G is best described not as a molecule of difference and defense, but rather as a molecule of tolerance (Carosella et al. 2008).

The expression of the HLA-G molecule by an allograft could encourage its tolerance in humans, as well as in mice. Indeed, a significant correlation exists between HLA-G expression and a lessening of rejection risks for heart, kidney, and liver allografts (reviewed in Rouas-Freiss et al. 2007). HLA-G seems to participate in the modulation of graft rejection by influencing all levels of the alloresponse (Carosella et al. 2008).

On a basic level, the discovery that HLA-G and regulatory T cells play an important role in immune privilege and specific graft tolerance suggests that the positive mechanisms of immune tolerance induction occur in numerous natural phenomena. As I will shortly show, immune inhibition by regulatory T cells, HLA-G, or certain cytokines occur in fact in fetomaternal tolerance, as well as in tolerance to tumors, to commensal microorganisms, etc. Yet before demonstrating the importance of this convergence, a question should be asked: all that I have just described concerns mammals, but what about immune tolerance to grafts in other species?

GRAFT TOLERANCE IN OTHER SPECIES

In plants and certain animals, tolerance to grafts is much more common than in mammals. This is the case with several colonial organisms, such as *Botryllus schlosseri*. In *Botryllus*, two colonies spreading underwater by asexual reproduction (budding) can run into each other. There are two possible outcomes: either they fuse, creating one colony, or they reject each other and a necrosis zone appears at the point of contact. Burnet had already noted that fusions between two allogenic colonies were possible and probably depended on the recognition of an allele common to both (Burnet 1971), which was amply confirmed by subsequent research (Stoner, Rinkevich, and Weissman 1999). *Botryllus* possesses a histocompatibility system (Scofield et al. 1982), very similar to the HLA system in humans, called the fusion histocompatibility complex (FuHC). Recently, the locus of this histocompatibility gene has been isolated and described (De Tomaso et al. 2005). If two *Botryllus* colonies have *one* of two common alleles, then this is enough for their encounter to lead to a fusion and not a rejection.

Fusions are equally common in sponges. Working on certain sponges, Hildemann and colleagues had noticed countless rejections and had deduced that fusion was a good criterion for genetic identity (Hildemann et al. 1979); in other words, if two sponges fused, one could conclude that they were identical products of an asexual reproduction. However, in a work on the *Ectyoplasia ferox* sponge, it was shown that in practically 50% of cases two allogenic sponges fused (Curtis, Kerr, and Knowlton 1982). According to these authors, diverse factors had to be taken into account, including the possibility that a single common allele was enough to allow fusion and the geographic proximity of the sponges. They noted that the ability to accept allogenic grafts varied widely among invertebrates, from 0.7% in the sea fan *Eunicella stricta* to 82% in the hydrozoa *Hydractinia echinata*.

Finally, in plants, of course, numerous grafts are possible and commonly practiced by botanists. Other grafts are produced spontaneously in nature, for instance, in the strawberry plant. Some "self-nonself" differentiation mechanisms could exist in plants, but they would come into play in reproduction and not in graft acceptance (Burnet 1971).

The mechanisms leading to graft rejection in different animal species are, as a whole, not well known. In any case, graft rejection implies innate immunity actors (including, for instance, Toll-like receptors), which is not surprising given that these mechanisms exist in most animal species, as we have seen.

Therefore, the self-nonself theory, which claims that all "nonself" is rejected, seems more or less acceptable depending on the species being studied; but taken as a whole, the theory is largely inadequate. This criticism, however, is aimed more at contemporary proponents of this theory than at its founder, Burnet. Indeed, it would be rather disingenuous to fault Burnet for having ignored phenomena of immune tolerance that are frequent in nature.

For his part, he claimed that his self-nonself theory only applied to higher vertebrates, or even just mammals. It would consequently be absurd to oppose his theory with the case of invertebrates or plants. Burnet strived to understand the acquisition, in the course of evolution, of this mechanism of rejection of the foreign. As we have seen, Burnet's thesis is that self-nonself differentiation exists in all organisms, but that *immune* recognition of the self is a recent product of evolution. On the other hand, contemporary immunologists, for their part, want to both offer one unified theory for all animals, including invertebrates, and uphold the self-nonself immunological theory. The latter is indeed applied to insects (Medzhitov and Janeway 2002), as well as to protochordates (Litman 2005) and even plants (Boller 1995). In doing so, immunologists today are formulating an erroneous assertion, since numerous organisms do not reject every foreign graft. It is therefore impossible to apply the self-nonself theory to all living species, and even to all metazoans. To avoid contradiction, contemporary proponents of the self-nonself theory would have to, as Burnet himself did, both limit the scope of their theory to only certain organisms (which considerably weakens it, given that research on innate immunity continues to show common points with so-called "adaptive" immunity), as well as explain the evolutionary history that leads to an immune rejection of nonself (which is rarely done).

Immunologists' focus on human transplantation and their scant attention to graft tolerance and rejection mechanisms in other organisms create difficulties. For instance, contemporary immunologists consider T cells to be central to immune response to an allogenic graft, something that is also suggested by these cells' presence on transplantation sites. However, the role played by innate immune cells, especially macrophages, in graft rejection in vertebrates could have probably been recognized much earlier if immunologists had focused more on organisms that are phylogenetically remote from man. Although it is obviously very difficult to take general lessons from the analyses offered here, perhaps it is still possible to suggest that immunology would benefit from concerning itself with all organisms, even when it comes to medical questions, such as that of organ transplantation.

In conclusion, the self-nonself theory becomes problematic as soon as one takes into account recent developments in mammal transplantation (notably in humans), and the mechanisms of graft tolerance and rejection in all living beings. I will now turn to two immune tolerance phenomena that strike me as quite important, namely fetomaternal tolerance and tolerance to microorganisms and macroorganisms: they most clearly show that the thesis according to which the organism rejects any exogenous entity is simply incorrect.

Fetomaternal Tolerance

WHY DOES THE MOTHER NOT TRIGGER AN IMMUNE
RESPONSE OF REJECTION AGAINST THE FETUS?

Pregnancy, because it can be described as the implantation of one tissue onto another, is similar to the process of a graft, to the point that it is generally understood as the most common natural graft (e.g., Burnet 1969; Brent 1997). The mother does not reject the fetus she is carrying even though it is genetically half different from herself, and thus even though it clearly belongs to "nonself." This is hardly a new observation, but it was for a long time explained as a process of immune isolation, on the model of immunoprivileged organs that I have just described: immunologists believed that the placenta was an impenetrable barrier to immune cells, therefore preserving the fetus from any rejection. This explanation was certainly in accordance with the self-nonself theory, or at least it was a way to save it; if the fetus were accessible to immune cells, then how would such a huge exception to the self-nonself rule be explained? In the course of the last decade, however, it has been shown that immune actors were quite present in the placenta, but that they did not trigger a rejection response against the fetus. Trophoblast cells, which form the external cell layers of the placenta, directly interact with components of the maternal immune system (Hunt 2006). These immune components include NK cells, macrophages, $CD4^+CD25^+$ regulatory T cells, as well as dendritic cells and $CD8^+$ T cells (these two populations being present in small number).

The absence of rejection of the fetus by the mother is an immunological enigma that has fascinated several generations of immunologists (Medawar 1953; Moffet-King 2002; Hunt 2006). This phenomenon probably results from several different mechanisms acquired throughout evolution. Nevertheless, in the course of the last fifteen years, two major mechanisms implied in fetomaternal tolerance have been demonstrated: the fetus's expression of the HLA-G molecule, and the inhibitory role played by regulatory T cells.

THE EXPRESSION OF THE HLA-G "TOLERANCE MOLECULE" BY THE FETUS

Trophoblast tissues do not express MHC class I molecules, so they cannot be the target of the mother's cytotoxic T cells. On the other hand, NK cells do respond to the absence of expression of HLA class I molecules, therefore they should trigger a destructive response against the fetus. And yet this is not the case. How then does the fetus manage to escape a destructive response even though it does not express "self" HLA molecules?

Such was the question that prompted the discovery, in the 1990s, of the HLA-G molecule's role in fetomaternal tolerance (Kovats 1990; Rouas-Freiss et al. 1997). This molecule expresses itself from the very first days of the sperm's fertilization of the egg. It allows the fertilized egg to be implanted in the maternal endometrium. In the physiological (i.e., healthy) state, the protein is expressed only in the placenta, the thymus, and the cornea. However, the messenger RNA that corresponds to HLA-G genes is expressed in most of the organism's cells. As a result, a stimulation, for instance by interleukin 10 (IL-10), can cause HLA-G expression, making the cell in question tolerant. The HLA-G molecule interacts with three receptors found in all cells of the immune system: B cells, T cells, NK cells, and antigen-presenting cells (including dendritic cells). HLA-G interacts with all of these cells by successively inhibiting NK activation, cytotoxic T cell activity, proliferative allogenic activity, antigen-presenting cell differentiation, and antibody production.

HLA-G's involvement in fetomaternal tolerance is based on three major observations: first, in repeated spontaneous abortions, HLA-G is absent in the placenta; second, in preeclampsia there is no HLA-G expression; third, in transgenic mice that express HLA-G, blocking it leads to spontaneous abortions (Carosella et al. 2008).

One remarkable discovery was the observation that HLA-G can lead other immunoregulatory entities to act in concert: HLA-G can cause the differentiation between certain subpopulations of $CD3^+CD4^{low}$ and $CD3^+CD8^{low}$ regulatory cells (regulatory cells that do not exist in the natural state, as opposed to so-called "natural" $CD4^+CD25^+Foxp3^+$ cells) (LeMaoult et al. 2004). This would explain why, although limited in number, HLA-G molecules can have strong inhibitory effects: by activating other regulatory mechanisms and rendering certain cells tolerogenic, they would ensure a general tolerance to the antigens concerned.

It is possible to conclude, then, that at the center of the histocompatibility system (HLA), which is traditionally defined as a system of inter-individual difference and defense, lies in reality one molecule (HLA-G) that induces tolerance, in the strict sense that I have defined, which is to say that it inhibits the activity of activating immune cells, either by direct interaction or by inducing the differentiation of regulatory cells.

THE ROLE OF REGULATORY T CELLS

The first demonstrations that CD4$^+$CD25$^+$ regulatory T cells may play a role in human pregnancy came from the observation of a regular increase in the number of these cells at each phase of pregnancy (Somerset et al. 2004; Mellor and Munn 2004). Shortly afterward, Varuna Aluvihare and collaborators demonstrated that such an accumulated presence of regulatory T cells during pregnancy was also seen in mice, suggesting that this mechanism may be found in all mammals (Aluvihare, Kallikourdis, and Betz 2004). Using the experimental potential of mice, they showed that the adoptive transfer of the mother's T cells without regulatory T cells leads to the failure of any semi-allogenic pregnancy, whereas the same manipulation would not change the outcome of a syngeneic pregnancy. This experiment shows that the mother's T cells represent a mortal threat to the fetus in the absence of maternal regulatory T cells.

The involvement of regulatory T cells (especially CD4$^+$CD25$^+$ Foxp3$^+$ cells) in fetomaternal tolerance is currently acknowledged in the case of mice and considered highly likely in human beings (Mellor and Munn 2004; Aluvihare and Betz 2006). Indeed, one notices in pregnant women an increase in the number of systemic and decidual regulatory T cells during the first six months. These CD4$^+$CD25$^+$ cells have similar structural and functional characteristics to mouse cells.

How do these cells intervene in fetomaternal tolerance? The question of knowing if regulatory T cells directly or indirectly inhibit effector T responses is not clear-cut. One possibility would be that the regulatory T cells cause a state of tolerance in antigen-presenting cells (APC), which then induce a state of tolerance in effector T cells

(CD4$^+$ and CD8$^+$ T cells). These indirect mechanisms could involve the local production of regulatory cytokines, such as interleukin 10 (IL-10) or TGF-β (*transforming growth factor-β*), that can themselves cause HLA-G expression.

In short, the mechanisms that ensure fetomaternal tolerance are diverse. These mechanisms, at the forefront of which are HLA-G and regulatory T cells, demonstrate that an organism can, in certain circumstances, radically and provisionally modify the repertoire of ligands against which it will not release an immune response of rejection. Long seen as the only major exception to the self-nonself theory, an exception which allegedly could be explained by "obvious evolutionary reasons," and which would be based on the fetus's inaccessibility to the mother's immune system, fetomaternal tolerance is in fact an active phenomenon, based on a specific inhibition and involving immune components that also play a role in other tolerance processes. Fetomaternal tolerance is thus no longer perceived as an isolated, enigmatic exception to self-nonself differentiation. It now appears as one of the main fields that could lead to a better understanding of the diversity of tolerance mechanisms.

BEYOND FETOMATERNAL TOLERANCE:
FETOMATERNAL CHIMERISM

One speaks of chimerism, as we have seen, when an organism contains cells bearing different genomes. Fetomaternal chimerism designates the fact that the mother conserves for a long period—perhaps even her entire life—cells from the infant she carried. It is a special, particularly long-lasting, form of immune tolerance. Fetal cells have been found in the mother up to twenty-seven years after pregnancy (Bianchi et al. 1996). In reality, this chimerism caused by pregnancy takes many forms: in addition to the mother keeping her child's cells, the child keeps the mother's cells, and even sometimes those of the grandmother. Fetomaternal chimerism occurs in all mammals: a mammal is thus always a chimera in the immunological sense of the word.

Some remarkable phenomena associated with fetomaternal chimerism have been observed recently. In particular, certain *pregnancy-associated progenitor cells* in the mother are capable of differentiating themselves in different lines (not just hematopoietic, but also epithelial and hepatic). These fetal cells may play a functional role, such as cell renewal, in the mother, for example, allowing tissue repair after a lesion (Khosrotehrani et al. 2004). The basis for this hypothesis is that one finds more chimeric cells originating in the fetus in injured zones than in uninjured ones. Faced with such recent, difficult-to-interpret experiments, it is important to remain cautious, but if they are confirmed, they would constitute an exciting example of immune tolerance of foreign components that acquire a *functional* role in the organism.

The body of analyses up to this point leads to two important conclusions concerning the self-nonself theory:

1. Graft acceptance seems to imply, in all species, the molecular resemblance of transplanted tissues or organs, but this resemblance does not depend on a strict genetic identity;
2. By active tolerance mechanisms, an organism's repertoire of non-immunogenic antigens can be enlarged considerably (as is the case with pregnancy).

These two conclusions have paved the way for the formulation of the objection I consider the most important with regard to the self-nonself theory—namely tolerance, by every multicellular organism, of countless microorganisms (and, sometimes, macroorganisms) within it.

Tolerance of Commensal and Symbiotic Microorganisms

THE IMPORTANCE OF THE IMMUNOLOGICAL PERSPECTIVE ON SYMBIOSIS

Symbiosis is a major biological process whose importance has been obscured for decades. Today, thanks to the work of many researchers (Margulis 1970, 1991; Sapp 1994; Margulis and Sagan 2002), it is increasingly accepted that symbiotic relationships are everywhere in nature (see Ernst Mayr's Foreword to Margulis and Sagan 2002). Symbiosis may be understood from numerous perspectives in biology, but when the phenomenon studied is intra-organism (endogenous) symbiosis, then the immunological perspective is crucial. Indeed, as immunology's aim is to understand what an organism rejects or accepts, it must explain this particular form of immune tolerance that manifests in a beneficial relationship between host and symbiont.

To be precise, the definition of symbiosis that I use here is the following: symbiosis is a lasting relationship between two organisms belonging to different species that is beneficial to one and neutral or beneficial to the other. This definition is in accordance with many contemporary symbiosis specialists (e.g., Hooper and Gordon 2001). I reject the definition adopted by certain people (e.g., McFall-Ngai 2002) who call symbiosis any lasting relationship between two different organisms whether it is beneficial or harmful. This definition is, in my view, far too expansive and does not allow for the description of characteristics unique to symbiosis.

The terms "commensalism" and "symbiosis" are both used to designate these relationships with microorganisms. In reality, it is not always easy to determine the frontier between commensalism, characterized by the absence of benefit for the two partners, and symbiosis, characterized by a benefit for one partner and at least the absence of harm for the other (and, in the best case, by mutual benefit).

An organism like the human being is in symbiosis with numerous bacteria, most often located at the organism's interfaces: on the skin, but also everywhere on epithelia: in lungs, intestine, vagina, etc. Most multicellular organisms contain a great number of symbiotic bacteria: plants (Kiers et al. 2003; Oldroyd et al. 2009), insects (Bourtzis and Miller 2008; for instance, it is estimated that up to 70% of arthropods could be infected by symbiotic *Wolbachia* bacteria: McFall-Ngai 2002), birds, mammals, etc. Given that bacterial symbiosis is quite studied in man (thanks to important technological advancements), I shall describe here the human example, but I shall also emphasize similarities to what is observed in most other multicellular organisms.

SYMBIOTIC BACTERIA IN THE INTESTINE

I begin with the study of the human intestine, the most important of the many and diverse symbiotic relationships the human organism has with bacteria. A human being contains a considerable number of symbiotic bacteria, at least ten times more than the number of cells bearing its "own" genome (Xu and Gordon 2003; Xu et al. 2004; Turnbaugh et al. 2007). The great majority of these are found in the intestine. These intestinal bacteria belong to between 500 and more than 1,000 different species. The human organism is critical to the survival of most of them (these bacteria thus cannot be cultured outside of the intestinal environment). Certain symbiotic microorganisms are prokaryotes, others are eukaryotes, but the former dominate (Round and Mazmanian 2009). Symbiotic intestinal bacteria are unique to each organism and, as a result, they constitute one of the best ways to individualize the organism. They vary according to where they are located, meaning that bacterial populations are not the same according to places in the intestine. In addition, they vary over time as a function of the environment (Hooper and Gordon 2001), mainly according to the host's diet, its encounters with other microorganisms (pathogenic or not), and the taking of antibiotics. Others, however, are very faithful to us. Already present bacteria, moreover, modify the environment (especially the nutritional one), and encourage the integration of new bacteria that are useful to the host, which in turn modify the environment, etc. (Hooper and Gordon 2001: 1117; Hooper et al. 2003).

The prevailing idea for a long time was that the immune system had no access to bacteria residing in the intestine: much like with immunoprivileged organs, this interpretation, in accordance with the self-nonself theory, consisted of explaining the tolerance to bacteria by the inability of the immune system to interact with them. As in all other cases where this idea has been advanced (immuno-privileged organs, fetomaternal tolerance, etc.), we now know that it is untrue. Admittedly, thanks to the mucus and glycocalyx lining and to the antimicrobial peptides that protect intestinal epithelial cells, bacteria found in the intestine's lumen largely do not pene-trate the intestine strictly speaking. Yet the intestinal immune sys-tem constantly interacts with these bacteria (Vaishnava et al. 2008; Round and Mazmanian 2009): bacterial antigens adhere to the intestinal mucus surface and interact with immune receptors car-ried by enterocytes, dendritic cells, and, especially, M cells (micro-fold cells) from Peyer's patches. Intestine's dendritic cells open narrow junctions between epithelial cells and send their dendrites outside the epithelium, which allows them to interact directly with bacteria present in the lumen (Rescigno et al. 2001). This discov-ery suggested that the mucus surface exposed to the intestine's microorganisms is much more important than what had long been believed. Bacterial antigens captured by antigen-presenting cells are then transported toward the mesenteric lymphatic nodes, where they are presented to T cells and they can stimulate antibody pro-duction by the resident B cells. The immune system does therefore have access to and can interact with antigens found in the intestine. Without this access several illnesses would affect the organism, in particular chronic inflammatory diseases of the intestine (inflammatory bowel disease, IBD), and most notably Crohn's disease. What the immune system does in the intestine, in fact, is to tolerate useful bacteria while destroying dangerous ones. Identifying the mechanisms by which commensal and symbiotic bacteria do not trigger an immune response of rejection is a rapidly expanding field of study (e.g., Coombes and Powrie 2008). Some of the main mechanisms are represented in figure 3.1. This intestinal immuno-regulation involves dendritic cells and regulatory T cells, especially

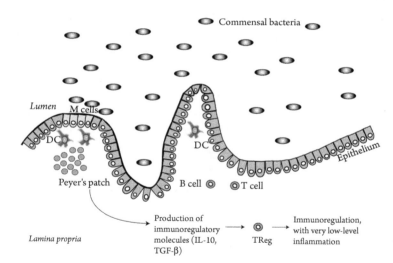

Figure 3.1. The intensive interactions between normal bacteria and the gut's immune system: Normal (commensal or symbiotic) bacteria are continuously and directly sensed by epithelial cells, microfold (M) cells, and intestinal dendritic cells (DCs). These bacteria induce the production of immunoregulatory molecules, in particular interleukin 10 (IL-10) and transforming growth factor β (TGF-β), which in turn activate regulatory T cells. This cascade results in an immunoregulatory response, characterized by a low-level of inflammation and an absence of destructive response against normal bacteria. These bacteria appear to be indispensable for the normal homeostasis in the gut.

CD4$^+$CD25$^+$Foxp3$^+$ cells. The first definitely play a fundamental role in presenting bacterial antigens to immune cells found in the intestine, but in a *steady state*, not an activated one. In other words, the conditioning of dendritic cells seems to be one of the main keys to inducing tolerance in the intestine (Coombes and Maloy 2007; Coombes and Powrie 2008). Furthermore, in the absence of regulatory T cells, the organism develops intestinal inflammatory diseases in response to bacterial antigens that are normally commensal. This observation has led to the demonstration of the immunoregulatory role of certain T cell subpopulations in the intestine, in conjunction with dendritic cell subsets (Coombes and Powrie 2008; Barnes and Powrie 2009). It appears that commensal and symbiotic bacteria in the intestine, which for all intents and purposes are "foreign," are nonetheless useful residents, stimulating tolerance processes that ensure their peaceful interaction with the host's immune system.

Moreover, these "foreign" bacteria are not only present inside the organism; they can also play a functional role and are sometimes even indispensable to the development, survival, or reproduction of the organism (one speaks of "obligatory symbiosis" to refer to a symbiosis indispensable to the host). This explains why the idea of "useful" bacteria is more and more frequently expressed nowadays. The intestine offers niches to gut bacteria, which allow them to feed and grow. In turn, these bacteria play diverse useful roles in the host. First, they enable the host to digest nutrients it otherwise could not (Xu and Gordon 2003). Moreover, the critical role of symbiotic bacteria in the host's development after birth has been shown both in mice and humans (Hooper 2004, 2005; Round and Mazmanian 2009; Eberl and Lochner 2009), especially in the normal development of gut-associated lymphoid tissue (GALT), that requires bacterial colonization (Hooper and Gordon 2001). A lot of information has been acquired through the study of *germ-free* mice: these individuals have a defective development, they possess an abnormal cellular composition in secondary lymphoid organs, an altered metabolism, a modified serologic composition, and changes in their cardiovascular physiology and neurophysiology (Macpherson and Harris 2004; Gordon et al. 2005; Round and Mazmanian 2009).

Even more surprisingly, symbiotic bacteria can play a crucial immune role in the organism. This phenomenon highlights how certain bacteria, far from being "enemies" that our immune system must fight, actually rank among its most essential components. Resident bacteria prevent other bacteria (which, for their part, would be harmful to the organism) from penetrating or moving into the organism. Many species of *Bacteroides* provoke the expression of antimicrobial molecules that bind directly to potentially pathogenic bacteria and eliminate them (Cash et al. 2006). In addition, many of these bacteria protect the host from strong intestinal inflammation (Kelly et al. 2004; Round and Mazmanian 2009), notably by inducing interleukin 10 (IL-10) production. More generally, the interaction of the host's TLRs with these commensal bacteria is critical to the regulation of inflammation and to the intestine's homeostasis (Rakoff-Nahoum et al. 2004). Moreover, bacteria located in the intestine are responsible for local immune responses, as well as responses in other areas of the organism (Noverr and Huffnagle 2004). Finally, it was demonstrated recently that in mice the gut microbiota acts as an adjuvant to induce immunity to the parasite *Toxoplasma gondii* (Benson et al. 2009), underlying the complex cooperation between host and symbiotic bacteria to get rid of harmful pathogens. Of course, pathologies may be associated with the presence of commensal bacteria: for example, certain cancers seem to develop only in the presence of commensal bacteria in the intestine. Nevertheless, this does not at all call into doubt their overall beneficial nature, which is even, in certain cases, essential.

In invertebrate species, symbiotic bacteria are often, as in man, essential to the host's survival, in particular with regard to nutrition (as is also the case for many plants, for instance), and also to development (McFall-Ngai 2002; Gilbert and Epel 2009). It has been shown, for example, that in the worm parasite *Asobara tabida*, the presence of symbiotic bacteria *Wolbachia* is crucial for oogenesis (Dedeine et al. 2001). Symbionts play a key role in the protection of their invertebrate hosts, notably in insects (Brownlie and Johnson 2009). Moreover, the immune response to gut bacteria in invertebrates is also highly regulated, resulting from a conjunction of positive and negative signals, as illustrated by recent studies in drosophila (Ha et al. 2009; Leulier and Royet 2009).

The tolerance of symbiotic bacteria in the intestine, ensuring functions vital to the host (nutrition, development, immunity) is therefore a very frequent, if not ubiquitous, phenomenon in multicellular organisms.

OTHER SYMBIOTIC AND COMMENSAL BACTERIA IN THE ORGANISM

Symbiotic bacteria are numerous in all the organism's epithelia, not just the intestine. For example, in mammals, of the ten "systems," eight (integumentary, digestive, respiratory, execratory, reproductive, immune, endocrine, circulatory) have a close and constant relationship with normal bacteria (the exceptions being, until proven otherwise, the nervous and skeletal systems) (McFall-Ngai 2002). Countless commensal or symbiotic bacteria thus exist on skin, in the vagina, in the lungs, etc. To take a single example, the respiratory tract in adult humans has a surface area of $70m^2$ and is constantly exposed to a huge number of airborne antigens; consequently, it is a highly regulated immunological area (Holt et al. 2008).

The organism, then, is fundamentally heterogeneous; it comprises different entities of different origins, including numerous bacteria that often play a critical role to its survival.

Tolerance and Pathology

The preceding analyses illustrate the importance of tolerance mechanisms to the functioning of a healthy organism. Yet these mechanisms can be hijacked or manipulated and may favor the development of a pathology, as happens in the case of pathogenic organisms and cancers.

HIJACKING OF TOLERANCE MECHANISMS BY PATHOGENS: THE CASE OF PARASITES

In the course of the arms race between pathogens and their hosts, it is common that the former (bacteria, viruses, parasites) have evolved strategies to avoid, or to manipulate, immune responses from the latter (Belkaid and Tarbell 2009). I am not going to go into detail here in the fascinating field of host-pathogen relationships, a field at the intersection of microbiology and ecology (Combes 2001; Wodarz 2006). I am going to focus instead on tolerance mechanisms in the strict sense of an immunoregulation, and I have chosen to look in detail only at parasites (helminths) one of whose remarkable characteristics is that they are often organisms of large size that can remain in their host for many years.

Most parasites display on their surface important quantities of antigens, many of which remain present in the host for a long time; however, in many cases, they do not trigger an immune response, or at least not an effector one. One very interesting example is that of schistosoma, parasites that infect 200 million people worldwide and kill 200,000 people per year. The schistosoma remains for a long time in its permanent host (an average of ten years) but does not trigger a strong inflammatory reaction. In the course of the last ten years, evidence has been accumulating of an induction of immune tolerance by the parasite. The schistosoma not only uses molecular mimicry but also synthesizes diverse cytokines that inhibit the immune system's lytic activity (Capron et al. 2005). The long-term acceptance of parasites like the schistosoma is thus increasingly understood today as the local induction of a microenvironment of immune tolerance.

Work done on several parasites, especially *Leishmania major* (which causes leishmaniasis), confirms these data. The presence of CD4⁺CD25⁺ regulatory T cells is a necessary condition for the persistence of *L. major* in its mouse host (Belkaid et al. 2002). Generalizing these investigations to other pathogens, it was shown that, especially with parasites, it is often the case that the pathogen's persistence corresponds to an "arrangement" between it and the host, made possible by immunoregulatory mechanisms, including regulatory T cells (Belkaid and Rouse 2005; Belkaid, Blank, and Suffia 2006).

TOLERANCE OF TUMORS

The HLA-G molecule, which plays an important role in the mother's acceptance of the fetus and could play a role in the tolerance of grafts, can also intervene in a process harmful to the organism: tumor growth (Paul et al. 1998). Even more fascinating, HLA-G can be involved in trogocytosis (from the Greek *trôgô*, "to gnaw") (Caumartin et al. 2007). Trogocytosis is a mechanism that transfers membrane fragments of one cell to another. Now, it has been shown that tumor cells expressing HLA-G1 molecules can transfer them to NK cells, making these NK cells for a short time stop being cytolytic and even causing them to behave like regulatory cells, allowing the tumor to not be destroyed by the immune system (Davis 2007).

The same thing has also been shown for T cells that, having acquired the HLA-G1 molecule, go from an effector state to a regulatory state (LeMaoult et al. 2007). The trogocytosis of HLA-G1 seems therefore to ensure a sort of emergency immune suppression, transitory in nature, that allows tissues that express HLA-G either normally (such as the fetus) or pathologically (such as tumors) to rapidly increase the number of regulatory T cells, inducing a tolerant state.

Furthermore, it has been demonstrated both in mice and humans that $CD4^+CD25^+Foxp3^+$ regulatory T cells accumulate in significant numbers at a tumor site. This observation, initially made in ovarian cancer in women, also revealed that regulatory T cells were specific to tumor antigens. In addition, the large presence of regulatory T cells correlates to reduced chances of survival. It appears that the tumor is thus capable of inducing a specific tolerance, allowing it to escape the effector immune response (Gajewski 2007).

Conclusion: Immune Tolerance and Refutation of the Self-Nonself Theory

A SIGNIFICANT MODIFICATION OF THE INTERPRETATION OF IMMUNE TOLERANCE

Recent data demonstrate a shift in the definition of tolerance, from the idea of a rare exception to the self-nonself rule to the idea of a frequent and essential regulation. Today, tolerance is considered as the result of a homeostatic equilibrium between activation and inhibition. Effector immune responses are accompanied by a strong inflammatory reaction, which can inflict considerable damage to the organism. The immune system has evolved in the dual direction of a strong inflammatory capability allowing the destruction of target elements, as well as a capacity to regulate this destructive response in order to avoid excessive damage. An efficient immune system, then, consists of effector and inhibiting mechanisms that allow the removal of damage-inducing antigens (pathogens, tumors, etc.) all while maintaining a state of tolerance with regard to functional entities, which can be endogenous as well as exogenous elements, like symbiotic bacteria.

IT IS FALSE THAT THE IMMUNE SYSTEM TRIGGERS AN EFFECTOR RESPONSE AGAINST ALL "NONSELF"

What I have just stated concerning grafts, fetomaternal tolerance, and the tolerance of microorganisms and macroorganisms calls into question the second claim of the self-nonself theory, according to which the immune system triggers a rejection response against all "nonself." In reality, each organism is populated by a vast number of foreign entities, mainly bacteria, but also viruses or even parasites. It is the flawed vision of the organism as "pure"—perfectly homogenous and endogenously constructed—that has led to the idea that an organism has to reject any "foreign" entity.

Faced with this recent data on immune tolerance, the self-nonself theory can only hold up at the cost of hypotheses that offer numerous exceptions to the self-nonself rule: after having systematically proposed explaining immune tolerance by immune ignorance, that is, the impossibility for the immune system's components to access the sites in question, proponents of the self-nonself theory put forth a series of ad hoc hypotheses to account for each of the observed tolerance phenomena. Thus, just as they speak in the case of a tumor of "altered self" (counting as nonself), so too do immunologists tend to say that the mother carrying the fetus possesses an "extended" self. The idea of a changing self can consequently just as easily justify the triggering of an immune response as its inhibition (i.e., immunogenicity just as easily as non-immunogenicity). All of these expressions only serve to conceal the absence of a precise definition of the "self" and lead to the tempting, but scientifically barren, idea that the "self" is ultimately just a synonym for "non-immunogenic." The only valid way to evaluate the relevance of the self-nonself theory is to give these terms one of the operative meanings I considered in the first chapter, and to determine whether the immune system indeed responds to nonself without responding to self. The body of data analyzed in this chapter illustrates the imprecision of this claim. It is therefore incorrect to assert that the exogenous is always and alone immunogenic: the criterion for the triggering of an immune response of rejection is not the external origin of the antigen.

CONCEPTUAL VAGUENESS OF THE SELF-NONSELF THEORY

The terms "self" and "nonself" have not received a precise definition in immunology. The most precise meaning, the one anchored in the individual's genome, does not prove the rule of immunological discrimination between self and nonself, since, as I have insisted, some tumor cells, for instance, belong to the "self" in the genetic sense, and yet they induce an immune response. Not one of the different meanings of these terms that I considered here meets the demands of the self-nonself theory. So, at present, "self" often really means "non-immunogenic," and "nonself" means "immunogenic," which, of course, dissolves self and nonself as a scientific theory. "Self" and "nonself" are, understandably, attractive terms that suggest immunology as the branch of biology that brings scientific discourse to bear on our identity. They have been kept more for the attraction they elicit, and because they unify the community of immunologists, than for what they could achieve as carriers of meaning. The word "self" is indeed at once very appealing, because it seems to tell us something about ourselves, human beings, and very flexible, capable of diverse definitions and interpretations that are not necessarily explicit, articulated, or coherent when taken together. In reality, the term's imprecision contributed to the fact that it imposed itself easily and lastingly: the "self" has only been maintained until today because no immunologist has tried to define it more exactly.

Anne-Marie Moulin (1990) and Alfred Tauber (1994) both speak of the immunological self as a "metaphor." By this, they mean that immunologists have imported the idea of reflexivity from philosophy and psychology, as in the idea that the organism must know its immune self and not respond to its own components. The use of this idea and of this term has led to countless shaky interpretations, notably the rather common one concerning the cognitive nature of "immune self" recognition (Tauber 1997; Atlan and Cohen 1998; Cohen 2000a). One widespread thesis among biologists and philosophers of biology is that metaphors and fuzzy terms are inevitable, and even heuristically useful, in science. In the field of immunology, Ilana Löwy (1991) and even Alfred Tauber, with Eileen Crist (Crist and Tauber 1999), support the idea that the strength of the self and nonself concepts resides precisely in their imprecision. The examination of the terms "self" and "nonself" suggests the illusions of such a position: to accept this metaphor's use as necessary and fruitful is to be condemned to never be able to distinguish between different metaphors and, above all, to not attempt to reduce the imprecision caused by the use of these metaphors. No science can be content with a radical conceptual vagueness, even if it is true that a science must always remain open to the contributions of external terminologies.

I believe I have demonstrated that the self-nonself theory did not allow for the full understanding of modern immunology's experimental data. Such a negative conclusion is not, however, sufficient. Indeed, it is now necessary to attempt to propose another immunological theory, anchored in another criterion of immunogenicity that would be consistent with the results of current immunology. This is what I shall try to do in the two following chapters.

Chapter 4

The Continuity Theory

Having illustrated the reasons for the dominance of the self-nonself theory in immunology during the past sixty years (chapter 2) and having shown why I believe that this theory is no longer valid (chapter 3), it is now necessary to propose a different theory that will, at best, replace that of the self and nonself, and, at least, I hope, reorient immunologists' thinking in a new direction. I shall try to do this here by offering my own theory, the *continuity theory*, which I first elaborated in 2004 following discussions with the immunologist Edgardo Carosella (see Pradeu and Carosella 2004, 2006a, 2006b). The continuity theory has earned the attention of a number of scientists (von Herrath and Homann 2008; Alizon and van Baalen 2008a, 2008b; Bourtzis and Miller 2008; Schmidt, Theopold, and Beckage 2008; Cooper 2010b; Parker et al. 2010); I hope, with this book, to convince others that it is an interesting path that merits further exploration. In this chapter, I explain the foundation and the detailed mechanisms of the continuity theory before proceeding to test the theory against existing theories in the subsequent chapter.

I begin here with the general idea of the theory in order to make what follows clearer to the reader. According to the continuity theory, *the triggering of an effector immune response is due to a strong modification of the antigenic patterns (ligands) with which the organism's immune receptors interact,* which is to say *a sudden appearance of antigenic patterns in the organism that differ strongly from those with which the immune system is continuously interacting.* However, I do not want this abbreviated version to be too hastily interpreted, for the theory is naturally more complex than it appears at first glance. Here are some important precisions, which will become clearer in the course of this chapter. First, the "continuity" I am referring to has both qualitative and quantitative dimensions. Second, the immune receptors concerned are those carried by T and B cells, of course, but also—and certainly even more important—by macrophages, dendritic cells, mast cells, granulocytes, etc., that is, cells belonging to the so-called "innate" immunity. Third, the "antigenic patterns" mentioned above are not necessarily "foreign" patterns: they can be either endogenous ("self") or exogenous ("nonself"). Indeed, according to the continuity theory, as we shall see in detail, it is not the foreignness of an antigenic pattern that defines immunogenicity. An immune response is not triggered because of the presence of "nonself," since, as I have shown in the preceding chapter, there are numerous cases where an immune response is triggered by "self" antigens, and, conversely, many situations where an immune response is not triggered despite the presence of "nonself" antigens. One of the continuity theory's more specific objectives is to explain immune responses to tumor antigens, which are endogenous antigens that nevertheless differ from normal antigens, and also tolerance to the countless symbiotic bacteria in the intestine and other interfaces, which, although they are "nonself" in origin, are not rejected. An intuitive way of expressing the idea of the continuity theory is to say that an immune response is triggered by the expression of *unusual* antigenic patterns, as immune receptors respond to ligands that are very different from those to which they have reacted up to that point. This does raise one potentially confusing point: it is not my intention to say that the immune system responds solely to *new* antigens: on the contrary, in species with immune "memory," antigens that have already been introduced in a given organism subsequently trigger stronger and quicker responses (this, of course, is the idea behind vaccination, as was explained in chapter 1). The criterion I am proposing, simply put, opposes the long-term presence of the antigen versus its sudden appearance, not the "never met" versus the "already met." Indeed, an antigen encountered for the second time is not "new" because it has already interacted in the past with certain immune receptors in the organism; but it is quite different from antigens that the immune system continuously interacts with at a constant degree of intensity. This simplified formulation, though far from being perfectly satisfying, creates a fairly clear image of the continuity theory.

One important question concerns the scope of the continuity theory. I shall try to show that my theory is valid for all organisms that possess immunity in the sense defined in chapter 1—thus at least for all multicellular organisms and, possibly, for unicellular organisms as well (see section "The Continuity Theory: From Multicellular to Unicellular Organisms" below). This leads me to emphasize what I consider as one of the main arguments in favor of my theory: it is a highly inclusive, unifying theory, meaning that it gathers under a unique explanation many different immune mechanisms, occurring in a number of different species.

I would now like to go into the detail of the continuity theory. What do the "continuity" and "discontinuity" I am referring to precisely mean? How are these "perceived" by the immune system? Where do these interactions between immune receptors and their ligands occur and under what conditions? In my formulation, I have proposed that only a strong antigenic discontinuity is immunogenic, since every expression of an unusual antigenic pattern does not necessarily trigger an immune response of rejection; but what is a "strong" discontinuity? My theory hinges upon the clarification of these points.

THE FOUNDATIONS OF THE CONTINUITY THEORY

Why Keep the Requirement of a Criterion of Immunogenicity?

The continuity theory aims to offer a *criterion of immunogenicity,* as does the self-nonself theory. A criterion of immunogenicity is a response to the question of when and under what conditions an immune response is triggered. It thus allows for differentiation between entities that trigger an immune response and those that do not. While I find the criterion offered by the self-nonself theory unsatisfactory because it contradicts numerous experimental data (as was shown in chapter 3), I do think that the need to propose a criterion of immunogenicity is still valid. The stance that it would be impossible, after the criticism of the self-nonself theory, to propose any other criterion of immunogenicity (Tauber 1999: 466–67; Vance 2000) should not be considered a priori as the right one. This is at best a makeshift solution, whose theoretical and therapeutic repercussions are considerable, for it demands adherence to the belief that immunology can be a strictly experimental discipline, lacking any precise theory. If the criterion of immunogenicity that I offer here proves inadequate, it will always be possible either to try to find a better one, or to adopt the solution of last resort that involves abandoning the search for such a criterion. But I believe that this moment has not come yet, and that is precisely why I offer here a detailed analysis of the continuity theory.

Putting Normal Autoreactivity and Tolerance at the Center of Immunity

The continuity theory takes as its points of departure the two main points of contention with the self-nonself theory that I have already put forth—normal autoreactivity and immune tolerance. It is thus with these two ideas that I shall begin the explanation of my theory.

I have shown that, contrary to what the self-nonself theory claims, the organism does react to its own components in a continuous way. An organism without normal autoimmunity does not survive. This autoimmunity not only involves lymphocytes but also innate immune cells, including phagocytic cells. Besides, this autoimmunity is put into place starting with the maturation of the immune system. Immune cells react with the organism's components maintaining a constant, average strength level of reactivity that guarantees immune homeostasis. This level of constantly maintained immune reactivity in the organism is measured in terms of affinity, specificity, and avidity of the biochemical interactions between receptors and ligands. It is therefore a precise, quantifiable piece of biological data. The continuity theory does not seek to understand why there are sometimes immune *reactions* and at other times not (recall that an immune reaction refers simply to the biochemical interaction occurring between an immune receptor and a molecular pattern). Rather, it aims to determine the conditions for the transition from the normal, constant level of autoreactivity to a stronger level of reactivity (always in terms of specificity, affinity, and avidity) capable of leading to an immune *response*, that is, of triggering *activating* mechanisms. An immune activation occurs only if there is a biochemical interaction of strong specificity, affinity, and/or avidity between immune receptors and ligands; the task then is to understand when these biochemical reactions switch from an average and constant level to a strong and unusual one. This question involves not only lymphocytes but also innate immune cells, notably macrophages and antigen-presenting cells (APCs), specifically dendritic cells. These cells interact continuously with the organism's normal components at an average intensity. What makes them interact with strong intensity to certain entities and trigger an immune response? The core assumption of the continuity theory is that molecular patterns that are constantly present in the organism arouse constant average interactions with immune receptors, while any strong expression of an unusual molecular pattern gives rise to a strong interaction, leading to an activating immune response.

The examination of immune tolerance in the previous chapter allowed me to specify the organism's normal components, those that are in constant interaction at an average level with its immune receptors. These normal components do not refer to endogenous antigens only, but to all antigens present in the organism at the moment it reaches immune maturity (plus possibly other antigens as well, as will be explained later). This is what the founding experiment of Billingham, Brent, and Medawar (1953) demonstrated. Tissues implanted early enough in the embryo or the newborn (depending on the species) can be tolerated indefinitely. What determines immunogenicity is not the endogenous nature of the antigen but rather its presence in the organism. This concept of antigenic presence is the foundation of the continuity theory.

With these two ideas of normal autoreactivity and tolerance in mind, I shall now detail the mechanisms of strong molecular difference, which lies at the heart of my theory.

ANTIGENIC DIFFERENCE AS THE TRIGGERING MECHANISM OF AN IMMUNE RESPONSE

The Fundamental Statement of the Continuity Theory

According to the self-nonself theory, the immune system perceives the difference between patterns expressed by self antigens and those expressed by nonself antigens. I believe that the self-nonself theory is right to explain immunogenicity by antigenic difference (i.e., molecular difference), but wrong to claim that what matters is knowing whether this difference is endogenous or exogenous in origin. The continuity theory asserts that immunogenicity is caused by molecular difference itself. It stands by structural (molecular) antigenic difference without insisting a priori that this difference would be immunogenic only if it were exogenous.

The continuity theory's central claim is that the triggering of an immune response is due to any strong discontinuity in the expression of antigenic patterns that the organism interacts with, which is to say the sudden appearance in the organism of antigenic patterns strongly different from those with which the immune system continuously (i.e., regularly)[1] interacts. It is thus a matter of a rupture in the continuity of molecular determinants interacting with immune cells. It is the appearance of a strongly different molecular pattern that explains the transition from an average, constant level of reactivity to a strong level of reactivity, this strong level causing an effector immune response. This antigenic discontinuity has qualitative and quantitative dimensions. Antigenic patterns that trigger an immune response may well be exogenous (alloantigens expressed on an organ transplant, bacterial or viral antigens, etc.) or endogenous (tumor patterns, patterns expressed by apoptotic cells, patterns recognized by regulatory cells, etc.) An immune response is triggered whenever immune receptors interact with antigens that strongly differ from those with which they normally interact—that is, those with which they repeatedly (regularly) interact and that remain the same or practically the same. This can happen when a pathogen penetrates an organism, when a tumor develops, or even when one of the organism's cells dies from apoptosis. In each instance, an effector immune response occurs, and the criterion for this immunogenicity is, in my theory, molecular difference rather than exogenicity. In other words, many (but not all, as I explain later) foreign patterns do trigger an immune response, but it is not because they are foreign, but because they are strongly different from the patterns with which immune receptors usually interact. Endogenous patterns do too trigger immune responses when they are strongly different from the patterns with which immune receptors usually interact. Figure 4.1 shows the very general principle of the continuity theory.

Before giving more details about the continuity theory, let me address one possible concern. Perhaps the terms "continuity" versus "discontinuity" will seem surprising to some readers, who might prefer the terms "usual" versus "unusual," or the "maintaining of molecular

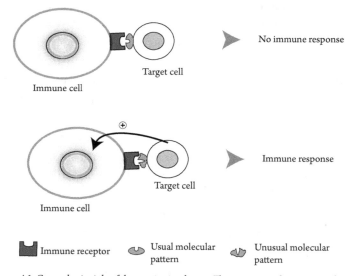

Figure 4.1. General principle of the continuity theory: The continuity theory states that the criterion of immunogenicity is not "nonself" vs. "self," nor "dangerous" vs. "benign," but molecular difference (or antigenic discontinuity) vs. molecular maintenance (or antigenic continuity). In other words, an immune response (i.e., the triggering of effector immune mechanisms) is due to a strong modification of the molecular patterns with which immune receptors interact. Therefore, the "target cell" leading to an immune response here can be a pathogenic bacterium or a cell infected by a virus, but also, for example, an apoptotic cell or a tumor cell (even though these are genetically "self" cells).

patterns" versus the "appearance of different molecular patterns." I must say that I am not strongly attached to the terms "continuity" and "discontinuity," and that I am open to any terminological suggestion. My point is simply that to say "what triggers an immune response is sudden molecular difference" constitutes a simple, experimentally adequate, and unifying explanation for immune phenomena.

Immune Receptors Involved

The continuity theory claims that an immune response is due to a strong discontinuity of the target's patterns with which the organism's immune receptors interact. But which immune receptors are involved? The answer is that all immune receptors are—not only those carried by B and T cells but also those found on the surface of "innate" immune cells: macrophages, monocytes, mast cells, dendritic cells, granulocytes, NK cells, etc. The activation of antigen-presenting cells is, I think, the most important problem any immunological theory must solve. Accordingly, the continuity theory concerns first and foremost APCs, which play a fundamental role in triggering immune responses. In particular, I believe that Toll-like receptors (TLRs) borne by APCs are an excellent example of immune receptors that are involved in the detection of antigenic difference (rather than "nonself"). Figure 4.2 shows two examples of immune activation as conceived of by the continuity theory: the activation of dendritic cells and that of T cells. (The activation of other immune cells, in particular phagocytic cells and regulatory T cells, is described below.)

Continuity's Point of Departure

The theory presented here hinges on the idea that an immune response is triggered when ligands appear that are strongly different from those with which the organism has regularly interacted until that point in time. Nevertheless, a question arises: at what point in time does this constant and repeated reactivity that antigenic discontinuity comes to modify begin? Antigenic continuity begins once immune cells are immunocompetent, that is, when their receptors effectively react with their ligands. Since immunocompetence may be acquired at various stages in the organism's development, it is necessary to say a few words here about the ontogeny of the immune system: that is, the setting up, in the course of the

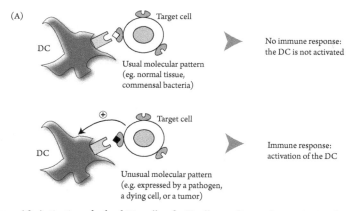

Figure 4.2. Activation of a dendritic cell and a T cell according to the continuity theory: (A) Dendritic cells (DCs) are distributed throughout the organism, interacting with all entities at hand. DCs bear receptors called patterns recognition receptors (PRRs), a large collection of receptors which includes, in particular, Toll-like receptors (TLRs). When the PRRs of DCs interact with usual molecular patterns (expressed by normal tissues or resident symbiotic bacteria, for instance), there is no molecular difference, and therefore no immune response is triggered. However, when they interact with unusual patterns (expressed, e.g., by transient pathogens like bacteria or viruses, or by dying body cells, or else by abnormal cells, including tumor cells), there is a molecular difference, and therefore an immune response is triggered. (B) T cells are activated by antigen-presenting cells (APCs) (typically, dendritic cells) that have themselves been already activated. If a dendritic cell presents at its surface, via major histocompatibility complex (MHC) class II molecules, usual peptides, then the T cell is not activated. If it presents unusual peptides and if it is indeed itself activated (thus expressing costimulatory molecules such as CD80 or CD86), then the T cell is activated. Naturally, not every T cell is activated in this case, but only those that bear receptors specific to the antigen presented by the dendritic cell.

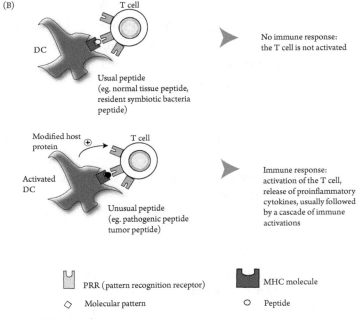

Figure 4.2. (Continued)

individual's development, of an effective immune system. I begin with mammals, in this case mice and human beings, whose immune systems are both particularly well known.

At birth, an organism moves suddenly from the sterile environment of the fetus to one rich in microorganisms. The development of the immune system begins during the early embryonic period. For example, mice have immunocompetent lymphocytes starting on the twelfth day after conception in an 18- to 21-day gestation period (Yokota et al. 2006). In humans as well, the first immunocompetent lymphocytes appear a few weeks after conception. Yet the immune system does require time to become completely mature. A fetus can trigger immune responses, but these are qualitatively and quantitatively different from those of adults. The immune system of a newborn is also quite different from that of an adult: in many aspects it is a diminished immune system, although it can still handle pathogenic threats (Levy 2007). In both mice and humans, certain components of innate immunity are effective from birth, while others do not activate until several weeks after birth. The situation is somewhat different when it comes to selected immune cells that undergo elimination if they carry strong autoreactive receptors, such as B cells, T cells, and NK cells. In fact, all cells carrying receptors that strongly recognize antigenic patterns introduced in the thymus are eliminated, whether they are endogenous or exogenous. By this deletion process, the immune system prevents the possibility for excessive immune responses to components present in the organism (and not to "self" components in the endogenous sense). The maturation of immune cells undergoing a selection (deletion) is a longer process than that of cells that are not selected. Thus, the newborn possesses a defective adaptive immunity, and its immunity relies mainly on innate mechanisms (Adkins, Leclerc, and Marshall-Clarke 2004). In mice, for instance, the lymphocytes' repertoire is not complete at birth. The thymus itself does not reach full development until the third or fourth week after birth. By contrast, it is fully developed at birth in the human being, but this does not mean that adaptive immune responses are already perfectly effective. Indeed, human beings possess mature lymphocytes several weeks after birth. The production of T cells actively continues until adolescence, when the thymus shrinks, reducing, but not completely stopping, T cell production.

The important point is that nonselected and selected cells start to interact with antigens present in the organism either before birth or a few weeks after it. They are, from this moment of maturity, positioned to interact in a continuous way and with a constant intensity to antigens present in the organism. As a result, the antigenic continuity that the continuity theory refers to begins with the maturation of the immune system.

In organisms other than mammals, immune maturation is often rapid. In animals with an exclusively "innate" immunity, immune receptors are to a large extent established at birth. In the fruit fly, most immune mechanisms are put into place during the larval stage (Lemaitre and Hoffmann 2007). In plants, the different immune mechanisms ("direct" and "indirect" pathways) are rapidly established and effective.

The Factors of Antigenic Discontinuity

My theory assumes that an immune response is due to a strong discontinuity of the ligands with which immune cells interact, not just any discontinuity. The continuity theory takes into account the following factors of discontinuity.

1. *Antigen quantities.* In most cases very small quantities of antigen will not provoke an immune response; if it does, this is very quickly interrupted (Kretschmer et al. 2005; Henrickson et al. 2008). Conversely, large quantities of antigen can paralyze the immune system, making it incapable of reacting. There is an antigenic discontinuity for the immune system only if the quantities of available antigens fall within a certain range (neither too weak nor too strong). Nevertheless, it is important to highlight that in the vast majority of cases introducing an antigen provokes an immune response. Most antigenic discontinuities are "perceived" by at least some of the immune components. The issue, however, is knowing when there is a move from an immune response at the level of one cell to a systemic immune response, involving a group of immune actors and ultimately leading to the target's destruction. In some cases, a cellular immune response begins, but regulatory immune

mechanisms shut it off, creating a state of tolerance to the antigen involved. Here again, the quantities of antigen presented is an important factor: as we shall see when speaking of tolerance by induction of continuity, small quantities of antigen presented repeatedly to immune cells can effect a state of active immune tolerance, and not a systemic immune response. It is important to understand, then, the circumstances where there is a transition from cellular immunity to global, systemic immunity.

2. *The speed of antigen appearance.* Just as the quantities of antigen play a role in triggering an immune response, so too does the speed at which the antigen appears. Some antigens that appear progressively in the organism do not provoke an immune response and can even trigger active tolerance mechanisms. On the other hand, antigens that appear suddenly in the organism will provoke a response.

To take both the quantity and speed of appearance into consideration, my theory could be formulated to say that what is important is not each of these aspects taken separately, but rather the relation between the two, that is, the dQ/dt, or the variation of antigen quantity with respect to time.[2]

3. *The degree of molecular difference.* The criterion suggested here lies in the molecular difference between constantly presented antigens and antigens appearing at a given moment in the organism. These differences carry degrees: if presented patterns are very similar to patterns that immune receptors have interacted with until that point in time, there will not be any immune response of rejection, since the antigenic discontinuity is not disruptive enough to be "perceived" by the immune system. The phenomenon of molecular mimicry that I have described in certain parasites demonstrates this, but so does histocompatibility in transplantation. This is a rather rare

occurrence in human beings, but, as I have pointed out, it is often observed in other species, such as colonial organisms. In *Botryllus*, for example, the continuity theory holds that the expression of several common antigenic patterns (linked to the possession of a single shared allele) suffices for acceptance and tolerance of a graft.

4. *The regularity of antigen presentation.* If an antigen is continuously present for a relatively long period of time in the organism and regularly interacts with its immune receptors, it can lead to tolerance instead of an immune response.

These last three points (speed of appearance, degree of molecular difference, and regularity of presentation) help to counter one possible objection: namely, that the continuity theory would make it impossible to understand an organism's normal endogenous modifications. As it matures and ages, an organism changes, its tissues are modified, and it undergoes genetic mutations, some of which have certain phenotypic consequences. The objection would thus be to say that my theory does not allow us to account for these normal endogenous antigenic changes in the organism. I would respond that antigens that appear after these changes are, in most cases, very similar to the organism's usual antigens, and that they appear slowly or are repeatedly presented to the organism's immune receptors. When this is not the case, then indeed an immune response targeting endogenous antigens is triggered. Here lies precisely one of the continuity theory's main arguments: strong antigenic discontinuities, *including those that are endogenous,* trigger an immune response. As I shall show in detail below, this notably occurs in the growth of tumors. Therefore, I believe the objection does not hold up: the continuity theory actually helps explain the non-immunogenic nature of weak antigenic changes, as well as the immunogenic nature of strong antigenic changes.

5. *The site of immune interaction.* Antigenic continuities are local; that is, the place in the organism where reactions between immune receptors and ligands are produced is an important element in determining if there will be an immune response of rejection. An antigen expressed by commensal bacteria in the intestine or on the skin and which is perfectly tolerated where it is normally found can trigger an immune response if it is introduced elsewhere in the organism.

Consequently, the continuity I speak of in my theory is spatiotemporal. It integrates the dimensions of space (the place of the immune interaction is important) and of time (the speed of the antigen's appearance and the regularity of immune receptors–antigenic patterns interactions are important). In this aspect, the continuity theory agrees with Zinkernagel who, in some recent ideas on the conceptualization of immune responses, has insisted on their spatial and temporal dimensions (Zinkernagel 2004), as well as with Grossman's and Paul's insistence on the context of the immune response (Grossman and Paul 2000).

Taken together, these five factors seem to open up the possibility of several states of immune tolerance, since only a strong molecular discontinuity causes an immune response of rejection. Consequently, I now turn to the role immune tolerance plays in the continuity theory's framework.

THE PRIMARY IMPORTANCE OF TOLERANCE IN THE CONTINUITY THEORY

Understanding the Ubiquity of Immune Tolerance Phenomena

Upon first reading, the central claim of the continuity theory could seem to suggest that once the period of immune maturity is reached, only that which is always, or almost always, alike (i.e., molecularly identical over time) is accepted by the immune system. The continuity theory, in other words, could give the false impression of simply proposing a principle of *conservation* of the organism's identity, with any change being a potential danger. I have already explained, though, that the continuity theory must not be interpreted this way: even when there is a certain antigenic difference, countless circumstances can lead to an absence of an immune response of rejection, such as small quantities of antigen, slow and/or repeated appearance of an antigen, etc. And as I shall show, this absence of an immune response of rejection is in many cases not due to an "immune ignorance" (absence of antigen–immune receptors interactions), but to an immune tolerance, or active immunoregulation. Indeed, the continuity theory places a central importance on tolerance phenomena. I believe that immune tolerance has not been properly integrated into recent immunological theorizations and philosophical reflections on immunology, and I hope that the continuity theory helps to better explain tolerance mechanisms and their ubiquity.

An organism is constantly exposed to and must constantly tolerate certain antigens, most notably oral (nutrition) and airborne (respiration) (Smith and Nagler-Anderson 2005). As I have shown in the previous chapter, an organism functions via "systems" (integumentary, digestive, respiratory, excretory, reproductive, immune, endocrine, circulatory, nervous, and skeletal) that are all, or almost all, (controlled) openings to the external environment. I also pointed out the huge presence of microorganisms, especially bacteria, that live in most of these "systems." This observation inverts the question of the self-nonself theory, no longer asking: "Why must all nonself be rejected?" Rather, since an organism is a system open to its environment (nutrition, respiration, etc.) and its immune system is thus constantly exposed to the presence of exogenous antigens sometimes vital to its survival, the main question is now: "How does this immune system distinguish, among so many antigens, those that are benign and even useful and those that are dangerous?"

Tolerance as Prevention against Immunity's Risks

The first step in my reasoning therefore consists of highlighting that immune tolerance is fundamental in the organism. The self-nonself theory thinks of the organism as largely closed to its environment, and then tries to lay out a list of exceptions to this rule of systematic nonself rejection. The continuity theory, in contrast, considers the organism as first and foremost tolerant to its environment (i.e., that it needs to be tolerant in order to develop and survive), but that it is capable of getting rid of entities (endogenous or exogenous) that could prove harmful or fatal, which are therefore exceptions to the rule of tolerance. In addition, organisms have evolved to avoid being destroyed by their own immune system, an evolution that involves normal autoreactivity rather than excludes it: the immune system must be tolerant to some antigens found at the periphery (especially, but not only, "self" antigens). One of the main underpinnings of the continuity theory is thus that immune tolerance comes first. Two ideas illustrate this claim: first, fetal or immediate postnatal tolerance; and second, the fact that several activating signals are required to effectively trigger an immune response of rejection.

THE REASONS FOR INITIAL IMMUNE TOLERANCE
ACCORDING TO THE CONTINUITY THEORY

As I have previously noted, Medawar and his colleagues demonstrated in 1953 that tissues implanted early enough (in the immediate postnatal period) in mice could be tolerated indefinitely (Billingham, Brent, and Medawar 1953). This result, amply confirmed by subsequent experiments, showed that the fetus and, in certain species, the newborn, are largely tolerant to exogenous antigens. How to account for this weak immunity at birth? The idea Burnet advanced, and which was constantly repeated until several years ago, was simply that the fetus's and newborn's (in certain species) immune systems were not yet fully mature, that is, were immunologically incompetent. We have seen, however, that the situation is far more complex: the newborn tolerates certain antigens, even while it triggers active immune responses of rejection with regard to other antigens. Within the continuity theory, the ontogeny of the immune system must be understood not as the passage from a deficient state of immune immaturity to a mature state where everything foreign is rejected, but rather as the transition from a strong tolerance to a weakened tolerance, during which immune receptors are generated on the basis of antigens (endogenous or exogenous) that are repeatedly or continuously present in the organism. This view explains both the initial immune tolerance shown in Medawar and colleagues' experiment, as well as states of immune tolerance that can be induced in an adult organism. I will now break down the first mechanism before moving on to the induction of tolerance in the adult (in the section "The Induction of Tolerance by Induction of Continuity" below).

In my theory, initial immune tolerance is not a deficiency (which would be corrected by subsequent maturation), it is necessary to the organism's construction, a construction which occurs in an environment and, in part, by integrating components from this environment. The initial tolerance is indispensable to the organism because, immediately after birth, it is exposed to a great number of microorganisms that it must not eliminate because they are vital to its nutrition, its immune protection against pathogens, and to the maturation of its immune system. The immune system has evolved to express a significant but selective state of tolerance that allows it not to discriminate between self and nonself but to diminish immune responses to useful microorganisms even as it destroys those that would cause damage. The continuity theory can explain this phenomenon, whereas the self-nonself theory cannot.

The first phenomenon to explain is immune tolerance to commensal or symbiotic bacteria in the neonate (Palmer et al. 2007). Of course, one answer to this question is to say that this tolerance results from a long coevolution process between hosts and bacteria. While I do acknowledge this background explanation, I am more interested here in analyzing the biochemical mechanisms involved in this tolerance.

Several processes facilitate the establishment of commensal or symbiotic bacteria in the organism. This is particularly the case with the reduced reactivity of the Toll-like receptor 2 (TLR2) in newborns, which is generally interpreted as one of the means to the normal establishment of beneficial intestinal bacteria in the host: this receptor, in effect, recognizes bacterial ligands like LPS (lipopolysaccharide) that is found both in pathogenic and commensal bacteria (Levy 2007). Moreover, newborns, both mouse and human, develop very limited inflammatory responses, no doubt because it is important to limit the risks of immune responses to microorganisms vital to the host's survival. By extension, numerous $CD4^+CD25^+Foxp3^+$ regulatory T cells exist in the newborn (Godfrey et al. 2005), and it is likely that these cells play an important role in the regulation of the organism's inflammatory responses, both to limit the risk of tissue destruction and to reduce immune responses to microorganisms that are useful to the host (Adkins, Leclerc, and Marshall-Clarke 2004). The fetus's and newborn's selective and relative state of tolerance thus does not reflect immune immaturity with regard to recognition of the self, but a state of partial openness to the environment. This openness is crucial to the organism and allows for the creation of its immune system on the basis of antigens that are repeatedly in contact with its immune receptors.

The second question is to determine whether this initial tolerance can be explained by the continuity theory. The newborn presents a high level of immune tolerance. It is thus useful to begin by investigating the specific conditions in which the newborn can respond to certain pathogens. Presumably, the newborn's immune system rejects entities that first are initially introduced in zones that are not interfaces (intestine, lung, etc.); second, are introduced in large quantities; and third, are introduced in a noncontinuous but rather discreet way, and not repeatedly. In contrast, exogenous entities, such as bacteria, which are introduced in small quantities into the organism and with which the immune system interacts repeatedly and progressively, are tolerated. This is typically the case with the bacterial colonization of the host that occurs in the intestine immediately after birth. Symbiotic bacteria clearly arrive there in large numbers, at the same time as pathogenic bacteria. Yet they are presented to the immune system in small quantities, continuously and frequently in a slow, repeated way. The intestinal immune system (in particular the mucus and the epithelial cells) acts as a filter or funnel: despite a massive presence of antigens, it only removes certain extracts (Rescigno et al. 2001), which allows it to detect very abnormal antigens and induce tolerance (by induction of continuity) at the same time. This explanation can be partially extended to the adult organism, as I will show in the section on induction of tolerance created by induction of continuity.

Although it initially fluctuates, the intestinal flora eventually stabilizes, that is, becomes faithful to the host. A continuity has been established and the resident symbiotic bacteria are, for the immune system, part of the organism's normal components.

THE IMPORTANCE OF MULTIPLE SIGNALS IN
TRIGGERING A LYMPHOCYTIC RESPONSE

Lymphocytes are at the center of the adaptive immune response. Once activated, they can also induce innate effector immune responses such as phagocytosis. Consequently, they are often described as the "conductors" of the immune response, especially "helper" CD4 T cells, which direct the activation of macrophages and B cells, among other components. Yet if lymphocytes may be said to organize the adaptive immune response, they do not trigger it. Lymphocyte activation depends on APCs, mainly dendritic cells. The continuity theory, following several immunologists that have played a major role in shedding light on the importance of innate immunity (especially Janeway 1989), emphasizes that immune responses are triggered by APCs (Lee and Iwasaki 2007). Several experiments have demonstrated that a lymphocyte is activated only when it receives multiple signals, each of them being a confirmation of the presence of "abnormal" antigens in the organism. At least two signals are required, as Bretscher and Cohn (1968, 1970) suggested, as well as Lafferty and Cunningham (1975).[3] In order to be activated, a T cell must be stimulated at the same time by its specific ligand carried by the antigen and by the "second signal" delivered by the antigen-presenting cell. As for B cells, they have to be stimulated both by their specific ligand and by an auxiliary CD4 T cell, called a *helper* 2 T cell (Th2). Since Bretscher and Cohn proposed their two-signal theory, it has been shown that in numerous cases, even more than two signals are required: for example, today it is accepted that three signals are necessary to activate a B cell (Lee and Iwasaki 2007), and a large body of research has been done on the notion of "immunological synapse," the contact zone between immune cell and antigen that would involve many receptors and co-receptors (Friedl, den Boer, and Gunzer 2005). Then one of the main prevention mechanisms against unwanted immune responses (i.e., those that could damage the organism) is the necessity for a lymphocyte to be stimulated by several signals to become activated. This requirement is not, of course, in any way specific to the continuity theory; however, it is in perfect accordance with its stance regarding the primary nature of immune tolerance. According to the continuity theory, a strong antigenic discontinuity is enough to activate components of "innate" immunity, especially APCs. For a lymphocytic effector immune response to be produced, though, multiple discontinuities are required, and usually several immune cells (innate and adaptive) must perceive the antigen they are interacting with as abnormal.

A dendritic cell that presents an antigen specifically to a lymphocyte, but which is itself in a steady state, activates this lymphocyte, which then starts to multiply ("clonal expansion") and yet this population of lymphocytes specific to the antigen is rapidly eliminated (Steinman, Hawiger, and Nussenzweig 2003; Belkaid and Oldenhove 2008). In other words, lymphocytes that would recognize their specific antigen without this antigen first also being recognized by the antigen-presenting cell are suppressed by an active immune tolerance mechanism.

This mechanism makes it possible to eliminate lymphocytes at the periphery that could damage the organism: remember that in the thymus many "self" antigens are not presented to T cells during their maturation, which means that each organism possesses strongly autoimmune T cells (see chapter 3). As a result, the organism must learn to tolerate these antigens, which will be presented to its immune cells at the periphery. If an antigen-presenting cell presents to a lymphocyte at the periphery its specific ligand, but this ligand corresponds to an entity that is continuously present in the organism (whether the entity is endogenous or exogenous), then the antigen-presenting cell, itself not activated, will cause the elimination, not the proliferation, of the lymphocyte (Steinman and Nussenzweig 2002).

All of this serves to illustrate that many components of the immune system prevent risks of damage that could be caused by other immune components.

FROM CELLULAR IMMUNE RESPONSE TO SYSTEMIC IMMUNE RESPONSE

One last form of immune tolerance occurs on a systemic level. To better understand when an entity is rejected by the immune system, it is necessary to distinguish three different levels of immunity. The first is that of *immune reaction*, that is, of the interaction between immune receptors and their ligands. The second is that of *cellular immune response*, which is the triggering of activating mechanisms in the cell that has interacted with its specific ligand. Some of these activating mechanisms correspond to the destruction of the target. As we have seen, this cellular immune response requires multiple activating signals. It is, however, completely possible that this effector response at the cellular level (which should lead to the target's destruction) is regulated by other activating responses, but this time inhibitory ones—and not at the level of antigen presentation, but later on, by inhibition of an effector response. For example, the HLA-G molecule or regulatory T cells can suppress effector immune responses. This leads to the third level, that of *systemic immune response*: an immune response leads to the rejection of the target unless it is downregulated at a systemic level (I will say more on the activation of regulatory T cells below).

Thus, at different immunological levels in the organism, many "confirmations" are required before a complete immune activation can occur. Tolerance is not a rare exception to the major mechanism of self-nonself differentiation. Instead, it is an active and necessary phenomenon for the organism's survival. The organism must prevent excessive damage to its normal components, whether these are endogenous or exogenous. The immune system does indeed trigger destructions that are vital to the organism's survival, but it carries the important risk of harming the organism. Therefore, it has evolved in the direction of preventing this risk with the help of active tolerance mechanisms. According to the continuity theory, the organism is from its birth very open to its environment, which allows it to integrate indispensable symbiotic microorganisms; once it reaches immune maturity, the organism is clearly less tolerant, since it rejects any entity that is very different from those with which its immune receptors repeatedly interact.

In addition, as I shall now show, the conditions for immunogenicity that lie at the heart of the continuity theory make possible, in the adult, the creation of a specific tolerance to certain antigens, that is, the establishment of a new antigenic continuity.

The Induction of Tolerance by Induction of Continuity

The induction of tolerance means the establishment, in an organism, of a specific tolerance to an exogenous antigen. It is at the very least difficult to explain this phenomenon using the self-nonself theory framework. In contrast, the continuity theory explains it using the notion of the induction of continuity, a process I shall now describe.

THE INDUCTION OF CONTINUITY
BY DENDRITIC CELLS

We have seen that dendritic cells have above all a "tolerogenic" activity. In other words, dendritic cells, which, as APCs, are usually considered the most effective for mounting an immune response, are also capable of inducing specific immune tolerance to the antigen they carry (Steinman, Hawiger, and Nussenzweig 2003). In particular, dendritic cells can induce specific tolerance in T cells to antigens they present in small quantities (Belkaid and Oldenhove

2008): This is notably the case with antigens from apoptotic cells (Steinman et al. 2000) or antigens from the environment that do not harm the host, or are useful to it (Steinman, Hawiger, and Nussenzweig 2003). In addition, allogenic T cells cultured with immature dendritic cells can become resistant to a new antigenic stimulation, even by mature dendritic cells (Jonuleit et al. 2000). A specific tolerance is established through an induction of continuity, that is, the repeated presentation of a same molecular pattern under non-immunogenic conditions (weak quantities, regular presentation, absence of inflammation, etc.) In conditions of nonactivation, the repetition of immune presentation to the lymphocyte leads to a lower (instead of greater) likelihood of being activated by the same antigen. Consequently, in addition to "immune memory," which has been known since the beginning of immunology, there also exists an inverse form of habituation: an antigen that is repeatedly presented to immune cells under non-immunogenic conditions will come to be tolerated by the immune system, which will subsequently have a very weak probability of triggering an immune response of rejection against this antigen. In other words, immune habituation seems to function both ways: on the one hand, an effector immune response leads (at least in certain organisms) to a stronger secondary response, and on the other hand the presentation of an antigen in non-immunogenic conditions leads to a greater tolerance.

In lymph nodes, dendritic cells continuously present antigens found in tissues to lymphocytes (von Andrian and Mempel 2003). They thus constantly induce a tolerance to endogenous antigens, as well as to antigens from the environment that are useful to the organism's proper functioning (antigens from nutrition, respiration, etc.) This is why it is possible to say of diverse antigens, endogenous or exogenous, that they induce tolerance by "continuing education" (Huang and MacPherson 2001).

Dendritic cells can be tolerogenic by themselves. Additionally, they can induce a tolerance by induction of regulatory T cells (Belkaid and Oldenhove 2008). Dendritic cells in steady state, and even, in some circumstances, mature dendritic cells that present nondestructive environmental antigens, can induce immune tolerance by inducing regulatory T cells (Smits et al. 2005). Furthermore, such tolerogenic activity might not be limited just to dendritic cells; it could exist in other innate immune cells, such as mast cells (Lu et al. 2006).

THE INDUCTION OF CONTINUITY
BY REGULATORY T CELLS

Regulatory T cells, especially CD4$^+$CD25$^+$Foxp3$^+$ T cells, can inhibit effector immune responses triggered against exogenous or endogenous antigens. In certain conditions, this induction of tolerance is the consequence of an induction of continuity. This is quite exactly what Waldmann and colleagues express by the concept of "negative vaccination" in the transplantation context (Waldmann et al. 2004, 2006). Indeed, it appears that in certain transplants, the organ or tissue graft itself creates a progressive tolerance to its antigens via the stimulation of regulatory T cells (Cobbold et al. 2006). The conditions for the induction of continuity by regulatory T cells are, again, small quantities of antigen and a progressive introduction (Chen et al. 2004; Kretschmer et al. 2005; Waldmann et al. 2006; Belkaid and Oldenhove 2008). In a remarkable experiment, Apostolou and von Boehmer (2004) introduced a small pump that regularly released small quantities of antigen under the skin of several mice the thymus of which had been removed, and which had no regulatory T cells. They noted that the helper CD4$^+$CD25$^-$ T cells of the mice could, thanks to this mechanism and in the absence of a thymus, become CD4$^+$CD25$^+$Foxp3$^+$ (and CTLA-4$^+$ and CD45Rb^{low}) regulatory T cells, that were in fact capable of regulating effector immune responses. Again, these results emphasize the existence of a tolerance-creating phenomenon (or "tolerization") via the repeated presentation of an antigen, as in the case of dendritic cells. Such tolerization is, here again, the opposite of immunization (i.e., the increasingly quick and effective rejection of an antigen that has been presented several times to the immune system).

EXPERIMENTAL DATA CORROBORATING TOLERIZATION VIA INDUCTION OF CONTINUITY

The research on induction of tolerance by dendritic cells and regulatory T cells is rapidly expanding (Coombes et al. 2007; Belkaid and Oldenhove 2008). All the answers are still far from being found. I think, however, that the mechanisms that will be discovered in the future will tend to confirm the notion of the induction of continuity. It seems likely that an induction of tolerance occurs via antigenic habituation, that is, a repeated presentation of an antigen in non-immunogenic conditions that leads to its prolonged tolerance. Even if the mechanisms of tolerance induction are not yet perfectly established, several phenomena already tend to confirm my interpretation.

First, it was demonstrated in the mid-1990s that a pregnant mother develops provisional tolerance to the father's antigens (Tafuri et al. 1995). This tolerance is specific, and it ceases after pregnancy. It manifests in the fact that during pregnancy the mother can, in mice, accept a graft that carries the father's antigens. We can hypothesize that this tolerance is due to an induction of continuity. The fetus begins its development in particular tolerogenic conditions, which include, in humans, HLA-G expression, presence of T_{Reg} cells, and absence of class I HLA molecules expression. Immune cells interact with these semi-allogenic antigens, which are initially in small quantities and encountered progressively. Some regulatory T cells specific to the father's antigens multiply at the site of the fetus's development, which creates a situation of induction of tolerance via induction of continuity. The immune system "gets used to" these exogenous antigens, which will be tolerated as long as they remain in the organism. Fetomaternal chimerism, which can last several years, reflects the fact that some of the infant's cells shift outside the placenta during pregnancy and remain in the mother in small quantities, interacting continuously but weakly with the mother's immune receptors. The mother can tolerate these antigens for a very long time or indefinitely; they could even play a functional role in the mother, as I pointed out in the previous chapter.

Another important example is "oral tolerance," when the immune system tolerates an exogenous antigen previously ingested by the organism (Chen et al. 1994, 1995; Dubois et al. 2005). Ingested antigens can lead to an antigenic continuity in the intestine that allows them to be tolerated. Antigens ingested in small quantities induce a tolerance by activating regulatory T cells (Dubois et al. 2005). Dendritic cells probably also play a role in oral tolerance, especially with low doses of fed antigens (Dubois et al. 2005). In particular, the key role played by plasmacytoid dendritic cells in oral tolerance is now established (Goubier et al. 2008). The induction of oral tolerance can be used to prevent allograft rejection, as well as autoimmune and non-autoimmune diseases, both in mice and humans (Mayer and Shao 2004).

The tolerance of symbiotic bacteria in the gut further illustrates the induction of tolerance by induction of continuity. I explained earlier the newborn's tolerance for certain bacteria that become commensal or symbiotic. But even though these bacteria are largely resident, important changes in their populations can be produced during the organism's life. The organism can become tolerant to certain "helpful" bacteria that it meets as adult. The first thing to emphasize is that the intestine is, more so than any other part of the organism, mostly tolerant: it is crucial to the organism's survival that the intestine does not trigger an immune response of rejection against the great majority of antigens that come from nutrition and resident bacterial antigens (Smith and Nagler-Anderson 2005). An entire series of tolerance mechanisms is involved: certain dendritic cells with a particular phenotype and that facilitate immune tolerance live in the gut associated lymphoid tissue (GALT); interactions of antigens with TLR are regulated by numerous cytokines, especially IL-2, IL-10, and TGF-β; the CTLA-4 receptor plays a tolerogenic role; there are multitudes of regulatory cells; etc. (Smith and Nagler-Anderson 2005; Kelly et al. 2005; see also figure 3.1 in the previous chapter). The organism, however, can trigger an immune response against pathogenic bacteria that penetrate the gut's barrier. The issue arises of knowing how the gut immune system is capable of making the distinction between pathogenic and useful bacteria. One response would be that pathogenic and useful bacteria carry different kinds of molecular patterns, but this is simply not true. Another explanation, rather easy, is to say that the immune system can distinguish what is "dangerous" from what is "harmless," but there is no way to support such a general hypothesis. A third answer that, though not completely satisfying, has some use is that organisms have simply evolved to not destroy certain "useful" bacteria. The argument is wholly valid when it refers to families of bacteria that are critical to survival, those the host needs in order to digest, regulate inflammation, and repair tissues (Rakoff-Nahoum et al. 2004; Kelly et al. 2004). Yet this argument needs to be completed by an explanation based on the induction of continuity. As much recent work has shown, there is a major difference between the way commensal bacteria and pathogenic bacteria are presented to intestinal immune cells (Strober 2009). Commensal bacteria are not so much "residents" (or "autochthonous" (Ley et al. 2006)) because they are tolerated as they are tolerated because they are residents. It is their ability to occupy intestinal niches that allows them to have a continuous presence and repeated interactions in small quantities with the gut immune system, and thus to be tolerated by it. The mucus that blankets the intestine's inner walls plays a fundamental role in this process. Most multicellular organisms, indeed, secrete mucus that allows them to "capture" useful microorganisms. To colonize the gut, microorganisms must be able to penetrate the mucus in which they will then live. A large number of antigens (for example undigested food particles, but also many microorganisms) simply do not penetrate the mucus. Most microorganisms that can are symbionts that find a niche where they can feed and reciprocally help the host, usually in digestion (Xu and Gordon 2003). These microorganisms are interacting with the immune system, but precisely because they are presented to it in small, ongoing quantities, they create a tolerance (Smith and Nagler-Anderson 2005; Strober 2009). Certain pathogens can, of course, penetrate the mucus, but if the resident bacteria do not eliminate them, they have to penetrate the host in large numbers in order to infect it, and therefore they elicit an immune response from the gut's immune system. The overwhelming majority of pathogenic bacteria (1) do not move into intestinal niches (which are occupied by commensal and symbiotic bacteria); (2) attempt to breach intestinal barriers in great numbers; (3) cause damage to the host, which produces pro-inflammatory cytokines that help activate immune cells.

The induction of tolerance by certain tumor cells or pathogens (notably parasites), and more generally the creation of tolerogenic microenvironments, can similarly be caused by an induction of antigenic continuity (see the explanation about the activation of regulatory T cells below).

The induction of continuity could also explain what is called "immune exhaustion." After extended contact with an antigen, T cells, in particular, can be exhausted; this is what happens in lymphocytic choriomeningitis virus (LCMV) infection (Gallimore et al. 1998). In such a situation, the immune system progressively ceases to respond to persistent antigens. *Clonal exhaustion by activation-induced cell death* describes immune exhaustion following the persistent presence of a pathogenic antigen. Repeated exposure to highly persistent antigens may be followed by clonal exhaustion (Yong et al. 2007).

Vaccination could appear to contradict my thesis that a small amount of antigen introduced into the organism creates tolerance, and not an immune response of rejection. It is true that vaccination introduces small quantities of antigen to cause an immune response that will produce a more rapid and effective response if there is a second encounter with this antigen. It is, however, important to point out that for the large majority of vaccinations, an adjuvant is combined with the antigen; this substance is added to the antigen to reinforce its immunogenic power. Without this adjuvant, there would not be an immune response to the small quantities of antigen presented in the vaccination process. As a result, vaccination does not invalidate my thesis of the induction of tolerance by the introduction of small quantities of antigen.

I have now laid out the essential principle behind the continuity theory. Since the initial motivation for proposing the continuity theory was the inadequacy of the self-nonself theory, it is now important to show how the former improves upon the latter.

DATA THAT THE CONTINUITY THEORY EXPLAINS BETTER THAN THE SELF-NONSELF THEORY

Normal Homeostasis, Autoreactivity, and Autoimmunity

The self-nonself theory cannot explain immune interactions with endogenous antigens, which are, however, indispensable to the organism's functioning, since immune cells need to be continuously stimulated by endogenous components. In contrast, normal autoreactivity is one of the foundations of the continuity theory: immune cells constantly and weakly react to the organism's normal components, whether these are endogenous or exogenous.

Not only can the self-nonself theory not explain endogenous immune *reactions*, it cannot explain the oft-observed effector immune *responses* to endogenous antigens either. Consider again the example of apoptotic cells, which are not "nonself": the phagocytic mechanism is the same for the phagocytosis of apoptotic cells as for a bacterium having penetrated the organism. Many phagocytic receptors—including scavenger receptor A (SRA), CD14 and CD36—interact with both endogenous and exogenous ligands (Taylor et al. 2005). Macrophages, in particular, respond to microorganisms, as well as to modified host cells (Savill et al. 2002; Stuart and Ezekowitz 2005; Taylor et al. 2005; Jeannin, Jaillon, and Delneste 2008; Green et al. 2009). Phagocytosis of dying cells is a true immune response, very similar to the phagocytosis of microorganisms or infected cells.[4] It is now well demonstrated that the phagocytosis of apoptotic cells is triggered by the interaction of receptors born by phagocytic cells with *modified* patterns expressed by these apoptotic cells. These modified patterns can be surface (CD14, scavenger receptors, etc.) or soluble (C-type lectin receptors, pentraxins, ficolin, etc.) molecules (Jeannin, Jaillon, and Delneste 2008). The continuity theory offers one unique explanation for the triggering of phagocytosis, whatever the target may be (a dying cell or a pathogen): in all cases, the immune response is due to a molecular discontinuity, with the target expressing molecular patterns that are different from those with which the immune receptors regularly interact (as shown in figure 4.3), typically extracellular nucleotides (Elliott 2009). Thus, the continuity theory unifies under a single explanatory mechanism what corresponded to two different "functions" of phagocytic cells in the framework of the self-nonself theory. More generally, the continuity theory returns to Metchnikoff's central idea that immunity is one of the main mechanisms creating homeostasis in the organism.

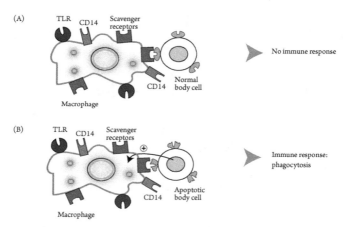

Figure 4.3. Phagocytosis according to the continuity theory: Macrophages are distributed throughout the organism, residing mainly in tissues and interacting with all entities at hand. They bear several pattern recognition receptors (PRRs), including Scavenger receptors, Toll-like receptors (TLRs), and CD14 (an endocytic receptor which binds, among other things, to lipopolysaccharide or LPS), as well as soluble bridging molecules or opsonins (not shown). (A) When the PRRs, and in particular the Scavenger receptors, of the macrophage interact with a normal body cell (bearing usual patterns), no immune response occurs. (B) When the PRRs, and in particular the Scavenger receptors, of the macrophage interact with a body cell dying by apoptosis (programmed cell death), an immune response occurs, namely the phagocytosis of the dying cell. In most (but not all) cases, the macrophage does not release proinflammatory cytokines and does not activate other immune components, which means that the cellular immune response of phagocytosis is not followed by a systemic immune response. (C) When the PRRs, and in particular the Scavenger receptors, of the macrophage interact with a body cell dying by necrosis (pathological cell death), an immune response occurs, namely the phagocytosis of the dying cell, the release of proinflammatory cytokines and the activation of other immune components, including dendritic cells (DCs) and natural killer (NK) cells. (D) When the PRRs, and in particular the Scavenger receptors, of the macrophage interact with the molecular patterns expressed by a bacterium, an immune response occurs, namely the phagocytosis of the bacterium, the release of proinflammatory cytokines and the activation of other immune components, including dendritic cells (DCs) and natural killer (NK) cells.

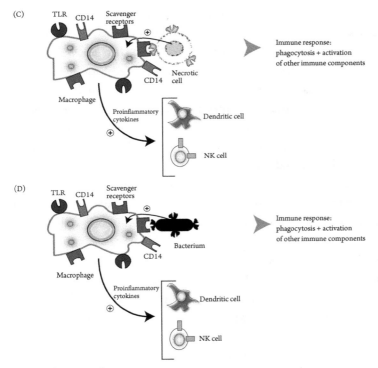

Figure 4.3. (Continued)

Activation of Regulatory T Cells

The activation of regulatory T cells cannot be explained within the self-nonself theory (see chapter 3). In contrast, the continuity theory offers a satisfying explanation for this activation. The key to understand regulatory T cell activation is to remember that the activation of adaptive immune cells is dependent upon the activation of APCs, in particular dendritic cells (DCs). In most cases, regulatory T cells are induced by APCs that are themselves not activated because they interact with usual antigens. The activation of regulatory T cells is then simply a direct consequence of the nonactivation of dendritic cells (steady state). It is well established that, in the great majority of cases, steady-state dendritic cells activate regulatory T cells (Jonuleit et al. 2000; Belkaid and Oldenhove 2008). In addition, it seems that in some tolerogenic conditions, mature dendritic cells can also induce regulatory T cells (Smits et al. 2005). Of course, it must also be remembered that regulatory T cells constitute one layer in what I described above as a multilayered immune tolerance system.

Two main categories of regulatory T cells (T_{Reg}) exist: "natural" and "induced" (or "adaptive") (Bluestone and Abbas 2003). "Natural" regulatory T cells are so named because they originate in the thymus, like other T cells, and have a regulatory phenotype from birth. They express CD4 and CD25 receptors, and the transcription factor Foxp3 starting in the thymic stage. At least some of these cells carry specific receptors for endogenous ligands. These cells are, contrary to other lymphocytes, selected for their *strong* ability to interact with endogenous ligands. "Induced" regulatory T cells, in contrast, are initially normal effector T cells that acquire the regulatory phenotype at the periphery, usually in the context of an infection. A transition in CD4 T cells from an effector phenotype to a regulatory one has been observed in many experiments (Chen et al. 2003; see also Bettelli, Oukka, and Kuchroo 2007).

In order to describe how the continuity theory conceives of the activation of regulatory T cells, four situations are to be examined (see figure 4.4):

1. *A steady-state dendritic cell presents a usual peptide.* In this case, the dendritic cell has interacted with usual peptides and is therefore tolerogenic; it stimulates regulatory T cells (in particular via the synthesis of immunoregulatory cytokines, such as IL-10 or TGF-β). In turn, these regulatory T cells prevent the development of a destructive immune response. In particular, this makes possible the prevention

of autoimmune diseases, a process which seems to involve mainly "natural" regulatory T cells, constantly stimulated at the periphery. This also makes possible immune tolerance in local microenvironments, as is best illustrated by immunoregulation in the gut. There, several APCs, including CD103⁺ dendritic cells, continuously present small quantities of food or flora antigens to lymphocytes and stimulate regulatory T cells (mostly "induced" ones), in particular via TGF-β and retinoic acid production (Coombes et al. 2007; Belkaid and Oldenhove 2008; Coombes and Powrie 2008; Barnes and Powrie 2009).

2. *A steady-state dendritic cell presents an unusual peptide.* In this case, even though the peptide is unusual, the dendritic cell

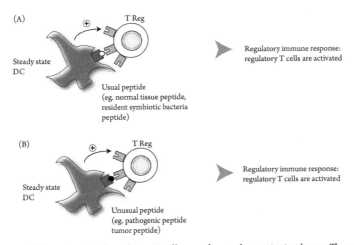

Figure 4.4. The activation of regulatory T cells according to the continuity theory: The activation of regulatory T cells is dependent upon the state of antigen-presenting cells (APCs), in particular dendritic cells. Four situations can be distinguished: (A) If a steady-state dendritic cell presents a usual peptide, then regulatory T cells are activated. (B) If a steady-state dendritic cell presents an unusual peptide, then regulatory T cells are activated. (C) If an activated dendritic cell presents a usual peptide, then regulatory T cells are activated. (D) If an activated dendritic cell presents an unusual peptide, then specific effector T cells are activated; when the quantities of antigen are low, some effector T cells progressively become "induced" regulatory T cells.

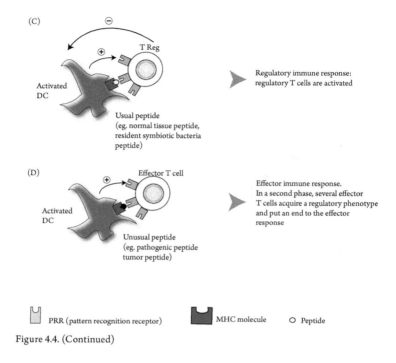

(C)

T Reg

Activated
DC

Usual peptide
(eg. normal tissue peptide,
resident symbiotic bacteria
peptide)

Regulatory immune response:
regulatory T cells are activated

(D)

Effector T cell

Activated
DC

Unusual peptide
(eg. pathogenic peptide
tumor peptide)

Effector immune response.
In a second phase, several effector
T cells acquire a regulatory phenotype
and put an end to the effector
response

PRR (pattern recognition receptor)   MHC molecule   ○ Peptide

Figure 4.4. (Continued)

has not "seen" it as unusual. Thus, if effector T cells interact specifically with this peptide, they will be eliminated (see above). In addition, some natural regulatory T cells will be activated, and some effector T cells will be converted to a regulatory phenotype. Indeed, many pathogens stimulate regulatory T cells via this mechanism, preventing a destructive immune response, as is well illustrated by the parasite *Leishmania major* (Suffia et al. 2006; Belkaid, Blank, and Suffia 2006; for a general perspective, see Pacholczyk et al. 2007). Tumors can sometimes do the same (Gajewski 2007). As we have seen in the section on the induction of continuity, microorganisms and tumors presented in small quantities may induce a new molecular continuity and will then not stimulate dendritic cells, and therefore will not be eliminated. Tolerance to new commensal or symbiotic microorganisms may be achieved through a similar route.

3. *An activated dendritic cell presents a usual peptide.* In this case, again, the regulatory T cells proliferate, they downregulate potentially reactive effector T cells (not eliminated in the thymus because the selection process to antigens present there is never complete), and they even downregulate the dendritic cells themselves (Shevach 2009). By this mechanism, regulatory T cells prevent autoimmune diseases (this implies mainly "natural" regulatory T cells) and detrimental immune responses to "useful" microorganisms, for example, in specific microenvironments as the gut (this implies mainly "induced" regulatory T cells), probably in an antigen-specific response (Smits et al. 2005).

4. *An activated dendritic cell presents an unusual peptide.* In this case, there is an antigenic discontinuity, which leads to the activation of the dendritic cell, so the dendritic cell stimulates effector T cells, not regulatory ones. No "natural" regulatory T cell bears receptors specific for this unusual peptide (because natural regulatory T cells are selected for their capacity to interact strongly with peptides present in the thymus). However, when the target (for example, a pathogen) has been eliminated to a large extent, much lower levels of antigens are presented by dendritic cells to lymphocytes, which is to say that the molecular discontinuity has become very weak. Progressively, some dendritic cells start to induce a regulatory phenotype in some of the effector $CD4^+$ T cells activated up to that point (weak quantities of antigen induce the multiplication of regulatory

T cells: Graca et al. 2005; Smith and Nagler-Anderson 2005; Belkaid and Oldenhove 2008). In other words, dendritic cells, now in a semi-mature or even a steady state, start to convert effector T cells into regulatory T cells. After this conversion, regulatory T cells contribute decisively to the elimination of effector B and T cells (but not memory B and T cells, which will survive the infection and be maintained in the organism). This would be the mechanism by which regulatory T cells prevent an excessive immune response, and in particular excessive damages to the host, in response to an infection (Suffia et al. 2006).

In conclusion, the continuity theory offers a coherent explanation for the activation of regulatory T cells: regulatory T cells are followers in the immune response; they do not "perceive" molecular difference, but rather, as key actors of immune tolerance, they are activated on the basis of an antigenic continuity, perceived whether by dendritic cells or by regulatory T cells themselves.

Immune Responses to Tumors

Can the immune system counter the development of a malignant tumor? The answer is clearly yes. Although certain tumors manage to escape from the immune system, the latter does detect and eliminate most nascent tumors, preventing them from even being physiologically detectable (Dunn et al. 2002, 2004). Yet in most cases, the tumor antigens are genetically "self" antigens, since their origin is endogenous: the expression of tumor antigens follows one or several genetic mutations, sometimes due to environmental influences. Although these are "self" antigens, in most cases an immune response is triggered against the tumor (Pardoll 2003). The self-nonself theory has a hard time explaining this process.

The first strategy it can use is to insist on the viral origin of certain cancers, which is a well-established fact for certain tumors. One speaks of "oncogenic" viruses to refer to these viruses; for example, the *Papillomavirus* can cause colon or uterine cancers. Nevertheless, not all cancers are viral in origin, and many are due to endogenous genetic mutations. The proponents of the self-nonself theory have thus coined the expression "altered self" to describe endogenous tumor antigens, claiming that the "altered self" functions like "nonself" (e.g., Houghton 1994). Yet with such a suggestion it becomes impossible to distinguish between normal modifications of the self (that are not immunogenic) and abnormal modifications of the self (that would correspond to the term "altered self" and be immunogenic). Thus, the term "altered self" serves in fact only to conceal an obvious insufficiency of the self-nonself theory, that is, its incapacity to explain immune responses to tumors.

By contrast, responses to tumors do find a coherent explanation with the continuity theory. Immune cells are activated, here again, by molecular difference. Tumor cells are not "nonself," but the patterns they express are different from those the immune receptors react with continuously. For the changes produced in tumor cells clearly differ from the changes found in the organism's normal cells: the genome of normal cells is mostly stable, whereas the tumor cells undergo multiple genetic alterations; the transcriptome is stable in normal cells, while tumor cells are characterized by a major epigenetic instability; no tissue invasion occurs with normal cells, but there is tissue invasion and formation of metastases with tumor cells; and finally, normal cells express cytokines and growth factors in stable way, while tumor cells express them abnormally (Pardoll 2003). Thus, tumor cells express molecular patterns which are very different from those expressed by the organism's normal cells; this difference constitutes an antigenic discontinuity, which is why tumor cells elicit an immune response.

The continuity theory thus constitutes a renewal of the idea of *immune surveillance*. This idea was first proposed by Burnet and, separately, by Lewis Thomas in 1957 (Burnet 1957; Thomas 1959; Burnet 1970). According to this hypothesis, the immune system constantly surveys the organism's tissues, and is in a position to detect and, in some cases, eliminate, tumors. This hypothesis, after having garnered much interest, was abandoned in the late 1970s. The 1980s and the 1990s, however, saw a resurgence of this immune surveillance hypothesis, largely due to the discovery of "natural killer" cells that certain researchers then saw as the main moderators of immune response to tumors and then to the work on "immunoediting" (Dunn et al. 2002). It is hard to hold on to the surveillance hypothesis using the self-nonself framework; by contrast, the continuity theory places at its core the idea that the immune system ensures a constant surveillance of any strong molecular modification. In doing so, it offers a relevant structure for explaining the idea of immune surveillance, but also extends this idea further than the sole surveillance of tumor cells. According to the continuity theory, the immune system constantly surveys all of the organism's tissues and responds to any strong change in antigens with which immune receptors interact. The immune system does not detect "nonself"; instead, it recognizes any strong molecular change, which is particularly the case in tumor development. This notion explains the phenomena of immune surveillance (or its equivalents) observed in both animals (Dunn et al. 2002) and plants (Chisholm et al. 2006).

Here again, the continuity theory unifies different findings under one explanation where the self-nonself theory had to settle for different explanations. As in the case of normal autoreactivity and autoimmunity, the continuity theory has a broader scope than the self-nonself theory and offers a unique explanatory mechanism for a large set of data, whereas the self-nonself theory must instead rely on a series of ad hoc exceptions and hypotheses.

Natural Killer Cell Activation

Natural killer (NK) cells play a major role in the elimination of intracellular pathogens, especially viruses. They also make essential contributions to the elimination of tumor cells. The common view is that NK cells do not respond to nonself antigens, but to the absence of self antigens created by the lack of expression of MHC class I molecules. This description of NK cell activation is already at odds with the self-nonself theory, which claims that all immune responses are due to the presence of "nonself" in the organism. But we can go still further: Some recent experiments have shown that NK cells respond not to the absence of MHC class I molecules, but to a modification of its expression (Gasser and Raulet 2006). These cells therefore seem to be responding to a molecular discontinuity caused by a molecular modification of the normal ligand of their receptors, that is, the MHC class I molecules. NK cells activation is one of the pieces of data that plainly illustrates the continuity theory's superiority over the self-nonself theory.

The Immunogenic Nature of Certain Mutations

Recently, it has been demonstrated that certain genetic mutations can induce an autoimmunity or anti-tumor immunity when they are expressed in inflammatory environments (Engelhorn et al. 2006). The existence of such immunogenic mutations seems to strongly corroborate the continuity theory's claim that it is molecular *change* that is the key to triggering an immune response.

The Alteration of Endogenous Components as Immunity Trigger

More generally, many species seem to be able to react to modifications of their endogenous components, which also bolsters the continuity theory. For instance, we have seen (chapter 1; see in particular figure 1.5) that plants often respond to their own "altered self," since the "indirect detection" pathway in plants is activated when endogenous "self" molecules have undergone structural modifications (DeYoung and Innes 2006).

Tolerance Phenomena

Whereas phenomena of immune tolerance to genetically foreign entities, though numerous, all are exceptions to the self-nonself theory, they are rather easily handled by the continuity theory. Countless nonpathogenic entities, especially symbiotic bacteria, are tolerated either because they are present at the moment of the immune system's maturation (and thus are part of the organism's normal components from the immune system's perspective), or because, though acquired in the course of the adult organism's life, they create a continuity by repeatedly interacting in small quantities with the organism, without triggering an inflammation. The continuity theory thus explains the organism's progressive tolerance of numerous useful, even indispensable, bacteria. Certain pathogens are also tolerated, which may be due to one of the four following reasons: they are not presented in the minimal activating conditions described above; they modify their antigenic patterns extremely rapidly; they inhibit the effector immune response; they induce a tolerance by inducing continuity (this last case occurs with certain viruses and parasites). In the same way, the induction of an antigenic continuity explains fetomaternal tolerance and chimerism.

In addition to the continuity theory's advantages that I have just described, I think that this theory helps explain all immune processes discussed up until this point in way that is at least as satisfying as the self-nonself theory. In other words, the self-nonself theory does not offer explanations of phenomena that the continuity theory could not explain, or would explain in a less satisfactory way. The rejection of microorganisms and macroorganisms is explained by the fact that they express different molecular patterns on their surface than those with which the organism's immune receptors have continuously interacted until that point in time. Antigenic difference likewise explains transplant rejections. An autograft and a graft between two identical twins are accepted because the antigens that these tissues or grafts carry on their surface are identical to those with which the recipient organism's immune receptors have continuously interacted until that point.

I deduce from all of these experimental arguments that the continuity theory explains certain immune mechanisms that the self-nonself theory fails to explain, and thus that contemporary immunology would have much to gain from adopting the continuity theory rather than continuing to develop the self-nonself theory as it has for more than a half-century.

THE CONTINUITY THEORY: FROM MULTICELLULAR TO UNICELLULAR ORGANISMS

The Continuity Theory in Multicellular Organisms

The continuity theory claims to apply to all multicellular organisms, not just to those endowed with adaptive immunity. I illustrate this thesis by taking two examples of multicellular organisms for which we have a sufficient amount of experimental data: plants and fruit flies (as an invertebrate example). To what extent is it possible to explain how their immune system works with the continuity theory?

In the plant's case, the continuity theory seems to apply very well. The immune receptors in the plant appear to recognize important modifications of its endogenous components, not "nonself."

This is called the "indirect recognition" process in plants (Innes 2011). More precisely, I will focus on three data to make the claim that the continuity theory applies to plants.

1. The existence of several normal autoimmunity mechanisms suggests that the immune response could be directed against modifications of endogenous components and not against exogenous components as such. The hypersensitive response (HR) is a form of programmed cell death, which is triggered against exogenous and endogenous elements. For example, it is triggered against certain pathogens upon recognition of pathogen-associated molecular patterns (PAMPs), but it also plays a fundamental role in the plant's normal construction and function, particularly in development and reproduction (Greenberg 1996). In addition, the existence of RNA silencing in plants has been observed, and there again is directed against both endogenous and exogenous elements (Chisholm et al. 2006).

2. The immune response is based on the recognition of disturbances. The most evidentiary support for the continuity theory comes from research conducted in the past ten years on *Arabidopsis* (Shao et al. 2003; Chisholm et al. 2006). This research seems to establish the existence of an immune surveillance in plants that consists of an interaction not with foreign elements but with strongly modified elements (Van der Biezen and Jones 1998). The point of departure for this research was that in many cases the plant's recognition of the pathogen seems to be indirect rather than direct (DeYoung and Innes 2006). In many instances, the plant effectively does not interact with the pathogenic patterns themselves, but with its own proteins and it responds to any strong disturbance in these proteins: "Rather than develop receptors for every possible effector, host plants have evolved mechanisms to monitor common host targets. By monitoring for perturbations,

R proteins indirectly detect the enzymatic activity of multiple effectors" (Chisholm et al. 2006).

By such an indirect mechanism, the plant can, from a limited number of receptors, resist a large number of different pathogens in a specific way. It is very likely that this mechanism of disturbances detection also applies to the recognition of "unwanted" endogenous elements. My hypothesis is that this mechanism is induced by a strong modification of the patterns that the plant's immune receptors interact with.

3. It is also highly likely that the LRR domains have immunoregulatory functions: they can probably inhibit an effector immune response. In certain cases, the deletion of an LRR domain in the potato leads to a more important hypersensitivity response which is, as we have seen, a form of programmed cell death. The indirect immune response in plants is activated against a complex made up of a protein targeted by the pathogenic effector and the amino-terminal domain of the NBS-LRR protein, as well as the NBS and LRR domains. The interactions of this complex are immunoregulatory, but when a pathogenic effector intervenes and targets a plant's protein, it disturbs the normal complex, and it is this disturbance that provokes the triggering of an immune response (DeYoung and Innes 2006).

It seems quite likely that the continuity theory also applies to the fruit fly (*Drosophila*). At the very least, it is certain the self-nonself theory does not account for the fruit fly's immunity. To begin with, the fruit fly is capable of autoreactivity and autoimmunity (Brennan and Anderson 2004; Lemaitre and Hoffmann 2007), like vertebrates. This autoimmunity takes several forms. Phagocytosis eliminates apoptotic cells, as well as exogenous components. The differentiation between the immunogenic and the non-immunogenic thus probably rests on molecular difference and not on the discrimination between "self" and "nonself." The fruit fly is also capable of regulating its immune responses or of triggering immune inhibition mechanisms (Zaidman-Remy et al. 2006; Bischoff et al. 2006). Such an inhibition allows the fruit fly to terminate the response and avoid excessive damages to itself (as in vertebrates), but it also helps it tolerate certain commensal bacteria. It is remarkable that the inhibition mechanism called PGRP-LB is activated by default in certain parts of the organism, notably the gut. Commensal bacteria in the gut, which are in small quantities and have a weak level of division, are perfectly tolerated by a process similar to the one I have described as induction of tolerance by induction of continuity. It is only when bacteria penetrate the gut in large numbers that they trigger an immune response (Zaidman-Remy et al. 2006; Leulier and Royet 2009). An immune surveillance exists in the fruit fly, allowing it to tolerate "useful" exogenous entities that are continually present in small amounts, and to respond to any endogenous or exogenous abnormal element (apoptotic cells, pathogens, etc.). Thus, the self-nonself theory does not account for current data on fruit fly immunity, whereas the continuity theory provides a plausible explanation for it.

The Continuity Theory in Unicellular Organisms

The question whether the continuity theory can be applied to unicellular organisms is a difficult one. As I showed in chapter 1, unicellulars do have an immune system, based on interference.

The continuity theory does account for immune interference, by applying its general principle to the genetic level (and not at the level of molecular patterns recognized by receptors located, for the most part, on the surface of immune cells). The genome has to face the potential threat of many viruses. The organism has to identify viral genetic sequences and prevent their integration into its own genome. According to the continuity theory, the mechanisms regulating the genome interact continuously with the genetic sequences that are present, and can trigger an interference response when unusual sequences show up in the genome. If the continuity theory is correct, then the exogenous or endogenous origin of these unusual sequences is not the decisive factor, and therefore we should expect in the future more discoveries on interference mechanisms targeting endogenous but abnormal sequences. Indeed, we are already seeing this today. There are two main forms of RNA interference in most eukaryotic organisms: one involves *small interfering RNAs* and principally ensures the elimination of viruses, and the other involves *micro RNAs* and chiefly ensures the regulation of endogenous gene expression. What about prokaryotes? The closest functional analogue of small interfering RNA seems to be the CRISPR system, despite some mechanistic differences (see chapter 1). The analogue for micro RNA is clearly established: prokaryotes have gene regulation mechanisms thanks to small antisense RNA, notably the Hfq protein for the presentation of small RNA and an RNAse E for degradation (Gottesman 2004). Thus, prokaryotes' immune system is most likely based on an interference mechanism that may be triggered by exogenous, as well as endogenous, elements. I suggest that even if the molecular mechanisms achieving gene "silencing" are not exactly the same, the recognition mechanism of the genes is the same: the target is not recognized as exogenous, but rather as composed of "abnormal" nucleic acids—that is, as different from the nucleic acids with which the effector proteins that trigger the Dicer proteins activation continuously interact ("Dicer" is the name of the protein that binds to double-stranded RNA and cuts them into pieces: Bernstein et al. 2001). As a result, I strongly believe that the continuity theory offers a satisfactory explication of interference-based immune mechanisms in general, and of its manifestation in unicellular organisms in particular.

CONCLUSION ON THE CONTINUITY THEORY

The core of the continuity theory is that it is not the exogenous nature of an antigen that explains its immunogenicity, but the fact that it expresses unusual molecular patterns in a strongly modified context. Overall, one of the main advantages of the continuity theory is that it gathers under a unique explanation different immune phenomena across phyla. In other words, it unifies phenomena that had been seen, up to now, as distinct, or even heterogeneous. Figure 4.5 sums up some of the most important of these phenomena.

The continuity theory is a scientific theory; it contains a certain number of hierarchical claims that can be experimentally tested, and it advances testable predictions. I will describe some of these here, because they are particularly significant. If the continuity theory is valid, then no immune response can occur without the appearance of unusual molecular patterns in the organism. This is a substantial claim to make because it excludes what certain immunologists consider as sufficient conditions for immunogenicity, such as inflammation and damages caused to the organism (I discuss these in depth in the next chapter). Similarly, the continuity theory predicts that no tumor triggers an immune response unless unusual antigens are expressed. It is even possible that certain autoimmune diseases may be triggered by abnormal mutations that lead to the expression of abnormal antigenic determinants. The continuity theory similarly predicts that very small quantities of antigen, introduced repeatedly into the organism, induce a tolerance rather than an

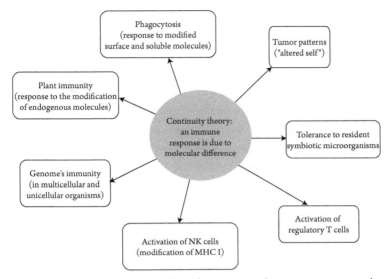

Figure 4.5. The continuity theory unifies under a unique explanation many immune phenomena: Several phenomena that are difficult to explain within the self-nonself framework and that have been described whether as exceptions (e.g., tolerance to symbiotic microorganisms) or using ad hoc hypotheses (e.g., tumor patterns defined as "altered self" and hence seen as immunogenic) are well explained by the continuity theory. Thus, this theory gathers under one and the same general explanation phenomena that were hitherto explained by different hypotheses.

immune response of rejection. This prediction leads to the hypothesis that the difference between tumors that are eliminated by the immune system and those that are not rests in the degree and speed of antigenic modifications (with regard to the usual patterns) expressed by these tumors. Likewise, the continuity theory predicts that the rapid appearance of a significant amount of antigen different from those with which the immune receptors have usually interacted until that point is immunogenic. For instance, the theory hypothesizes that numerous future experiments will show that some endogenous mutations are immunogenic if they lead to an unusual and partially stable phenotype[5]; by the same token, it theorizes that future experiments will show that at the source of normal autoimmune responses, in particular phagocytosis, there are molecular modifications on cell surfaces.

Thus, some experimental data could in the future corroborate or invalidate the continuity theory. Until then, I believe that the continuity theory provides a compelling alternative to, and even replacement for, the self-nonself theory.

To finish, I will sketch out some possible paths for therapeutic research that would be greatly aided by the continuity theory. My theory suggests developing research programs on the induction of immune tolerance. These could particularly involve commensal and symbiotic bacteria: it is time to stop considering every bacteria as an enemy, and instead look to the development of a "useful" bacterial population. I am especially thinking of probiotic strategies, which involve first and foremost intestinal flora (Noverr and Huffnagle 2004). Stimulating certain resident bacteria can help eliminate certain diseases (Di Giacinto et al. 2005; Benson et al. 2009). On the one hand, the elimination of all bacteria leads to the creation of more virulent strains, and on the other hand, organisms, notably humans, that are born in a sterile or nearly sterile environment have defective immune systems. The induction of tolerance could also play an important role in the field of transplantation, though this is a difficult issue. The continuity theory suggests that some "habituation" strategies to donor antigens before performing the transplant are possible and would allow a better tolerance of the graft. Understanding the mechanisms of tolerance induction would also allow, in certain cases, for the prevention of such an induction. For instance, the continuity theory postulates that one of the reasons for the development of tumors is their capacity to induce a tolerance in the immune system by stimulating the regulatory T cells or expressing the HLA-G molecule. Preventing the induction of continuity, by stimulating APCs (dendritic cells, in particular), could be the solution for breaking tolerance to these tumors. To be sure, these are only modest ideas. I do believe, however, that the continuity theory could lead to the creation of new and productive therapeutic research programs.

Having presented the continuity theory, I now examine how it compares to other existing theories in immunology.

Chapter 5

Comparing the Continuity Theory to Other Immunological Theories

This chapter consists of a systematic comparison between the continuity theory and several competing ones that already exist, from the self-nonself theory to those theories that have been developed as a critical response to it. My goal is to not only clarify the advantages of the continuity theory further in making these comparisons but also to point out those elements put forth by established theories that directly inspired the formulation of the continuity theory.

COMPARISON WITH THE SELF-NONSELF THEORY

Because the continuity theory was constructed as an alternative to the self-nonself theory, much has already been said, in the previous chapters, about the differences between these two theories. Yet one possible objection remains, by which the conceptual flimsiness of "self" and "nonself" comes to the fore here again. Could not what I call "continuity" and "discontinuity" be seen as, in fact, synonymous with the words "self" and "nonself"? In other words, how is the continuity theory truly different from a simple reformulation of the self-nonself theory? Here I show why I reject this objection.

If "Self" Is a Synonym for "Organism," then the Continuity Theory Must be a Restatement of the Self Theory?

I will begin with looking at one way that the self theory and the continuity theory may be misconstrued as the same thing. The continuity theory claims that some entities are part of the organism, even though they originate outside the organism, because they are in continuous interaction at an average level with the organism's immune receptors. The following objection arises: What prevents us from calling this organism the "self," a self that would include the "foreign" entities that it tolerates? For example, could we say that commensal and symbiotic bacteria make up part of the organism's "self," as do tissues implanted very early in the mouse neonate, which are tolerated throughout the organism's life? This terminology would account for the non-immunogenic nature of these bacteria while preserving the self-nonself theoretical framework. Conversely, any entity that would break the continuity of the ligands with which immune receptors interact could be called "foreign." All of this amounts to the idea that the "self" of an organism includes all the foreign entities that it does not reject immunologically.

In one sense, this is the thesis that I am defending, as the next chapter will show in greater detail: The organism must be defined based on the continuous, moderate level of its immune interactions. However, it is illegitimate to infer from this idea that the continuity theory would be equivalent to the self theory, for a reason that we have already pointed out. Using the term "self" as a synonym for "organism" in this instance is actually referring to the "non-immunogenic," which means that this logic presupposes what is in question. The terms "self" and "nonself" become, in effect, simply descriptive, and they cannot offer a basis for a true scientific *explanation*. If we call anything that triggers an immune response of rejection "nonself" while claiming that any immune response of rejection is explained by the penetration of "nonself" in the organism, then we find ourselves back in the circle we have already met where "self" and "nonself" cease to play any explanatory role (they become purely descriptive terms). Thus, to claim that the "self" and the organism are one and the same is fine, but this claim yields nothing in terms of a criterion of immunogenicity, which is to say an explanation of the triggering of an immune response—which is, by contrast, the very objective of the continuity theory.

Does the Self Theory Already Define "Nonself" as Antigenic Discontinuity?

There is a second way of asserting that the self-nonself theory and the continuity theory are the same that is much more compelling than the first. It consists of the claim that theoreticians of self and nonself use the term "nonself" to designate precisely a molecular discontinuity at the level of the ligands with which immune receptors interact. In other words, the continuity theory would only have substituted one term ("molecular discontinuity") for another ("nonself"). There is no easy way to evaluate the relevance of this objection, since it is rather impossible to study all uses of the word "nonself" in the immunological literature in order to determine whether it *may* be signifying "molecular discontinuity." Nevertheless, at least two texts that deal with the immune self seem to encourage the support of this thesis, according to which, in the self-nonself theory, "nonself" simply refers to "molecular discontinuity." The first is Philippe Kourilsky and Jean-Michel Claverie's description of the "peptidic self model" (Kourilsky and Claverie 1986). For example, in a 1990 summary Claverie writes: "By a simple opposition to the self, the only common denominator to the nonself is its transitory, episodic nature, its sudden confrontation with a *mature* immune system" (Claverie 1990: 36).[1]

The second text comes from Burnet himself, which seems to further jeopardize the strict distinction I have established between the continuity theory and the self-nonself theory. At the end of his scientific career, and in particular in his book *Self and Not-Self*, which appeared in 1969, Burnet seems to conceive of the "nonself" as a molecular discontinuity. He asserts that an antigenic determinant is in fact "foreign" when it is *absent*, and it belongs to the "self" when it is present from the embryonic stage onward and persists in the organism: "Recognition that an antigenic determinant is foreign requires that it shall not have been present in the body during embryonic life. Conversely, any foreign cells introduced early enough in life will be accepted as if they were the body's own cells for as long as they persist" (Burnet 1969: 25).

Such a statement conforms to his "immune surveillance" hypothesis, which he was also developing at the time. He writes specifically: "Discussion of the vertebrate, and specifically the mammalian, immune mechanism leads almost inevitably to the conclusion that it is more basically concerned with the control of tissue integrity and reaction against recognized anomaly in tissues than in defence against microorganisms and the production of antibody" (Burnet 1969: 22–23).

Of course, one could insist on the fact that Burnet is far from always having made such remarks. However, the objection is well founded if it is rephrased in such a manner: Isn't the continuity theory a simple reformulation of the self-nonself theory as Burnet understood it at the end of his scientific career, as well as Claverie and Kourilsky in the 1980s?

To respond to this objection, I return to the heart of the most significant modification proposed by the continuity theory. The theory claims that an immune response is due to a molecular difference in the targets of immune receptors, rather than the exogenous ("foreign") nature of this difference. This molecular difference must be understood with regard to the construction of the organism throughout its lifespan. In contrast, the self-nonself theory states that the "self" may be alterable, but only during the period of immune immaturity. For instance, in Medawar's experiment, anything that differs from the mouse's own tissues and those exogenous tissues implanted early on may be deemed "foreign." Accordingly, "foreign" would be a (poorly chosen) term that would indeed serve to designate that which differs from a molecular identity capable of accommodating exogenous elements, but this reference identity would be defined once and for all at birth or shortly after. The organism's identity as it is conceived under the continuity theory is quite different. As I have already shown using the notion of the induction of tolerance by induction of continuity, the standard identity in relation to which there can be antigenic discontinuity is constructed, and thus may vary, throughout the organism's entire life. Of course, this is simply another way of restating the observation that whereas immune tolerance is limited to the organism's immature phase in the self-nonself theory, it can be induced throughout a lifetime in the continuity theory. I may conclude, then, that the continuity theory is not just a reformulation of the self-nonself theory.

These remarks close the comparison between the continuity theory and the self-nonself theory. I cannot, however, remain content with this comparison, since I am certainly neither the first to find the self-nonself theory insufficient, nor to offer an amended or alternate theory. Several competing theories exist today that are useful to analyze here. My aim is both to compare the continuity theory with these other theories, and to point out those instances in which the continuity theory elaborates on and borrows from ideas that were originally expressed within these theories.

COMPARISON WITH SYSTEMIC THEORIES OF IMMUNITY

The first important critique of the self-nonself theory was suggested in the mid-1970s as a consequence of "systemic" immunological theories, or theories of the immune "network." Jerne's 1974 seminal article first established this approach (Jerne 1974). In it, he proposed a definition of the immune system as self-centered and fundamentally autoreactive, that is, reacting most of the time to the "self" rather than to the "nonself." Jerne's systemic vision has proved crucial because it has framed all competing views of the self-nonself theory until today, including autopoiesis (Humberto Maturana, Francisco Varela, Antonio Coutinho), self-organization (in particular Irun Cohen and Henri Atlan), and finally, at least in part, the danger theory (Polly Matzinger). The general idea held by all these partisans of a systemic vision of immunity is that the immune system does not discriminate between self and nonself because it only ever deals with the "self" (Tauber 1999). The paradox is that Jerne's systemic vision, which established a framework for all criticisms of the self-nonself theory from the 1970s on, was not itself presented as a critique of the self-nonself theory. Jerne never considered himself an opponent of self-nonself. For instance he writes in 1984: "Self-nonself discrimination may well be the most important problem facing the evolution of the immune system" (Jerne 1984: 17). I will now analyze in depth the systemic view before moving on to illustrate how the continuity theory both draws upon and differs from it.

Jerne's Definition of the Immune System as a Self-Centered Network

Jerne, a 1984 Nobel laureate, left his mark on immunology mainly through his highly theoretical way of thinking, deeply influenced by philosophy (Moulin 1991). In the 1970s, he proposed a systemic vision of immunity that intrigued many of his peers and had considerable influence on theoretical immunology up until the 1990s.

The 1974 article presents itself as a theoretical anticipation: Jerne is trying to predict what the main lines of inquiry will be in the field of immunology for the next twenty years. His main thesis is that immunology will go beyond the clonal selection theory (which, he rightly recalls, he helped create) by integrating a larger perspective that conceives of the immune system as a "network." This new way of thinking about the immune system has two aspects. First, it consists of moving from the idea that immunity plays out at the level of individual cells or even antibodies to the idea that it occurs in fact due to multiple interactions between many different cells. Second, it implies that the immune system is principally oriented inward (toward the "self"), because the organism's antibodies are initially produced in the context of an absence of "nonself," and later on they react continuously with the organism's normal components even though they simultaneously undergo a regulation that prevents them from destroying the organism. Jerne expresses three major ideas that are therefore important to analyze here further.

AN IMPORTANT BUT IMPRECISE CONTRIBUTION CONCERNING IMMUNE AUTOREACTIVITY

The first and most fundamental of the ideas Jerne proposed is that the immune system is essentially self-centered, that is, only ever deals with the "self." What exactly does this mean? Jerne explains it in 1974, by relying on the concept of "eigen-value" (or "eigen-behavior," an idea Heinz von Förster used, which inspired the majority of theorists of self-organization), used at that time in computer science and cybernetics: Any antibody capable of recognizing antigens is itself antigenic, that is, it can trigger the production of specific antibodies. We have in our organism antibodies that are recognized by some auto-antibodies, which are in turn recognized by other auto-auto-antibodies, and so on infinitely. What then follows is interplay between stimulations and inhibitions, which makes the immune system react continuously to its own components but without provoking the organism's self-destruction. Moreover, from a systemic point of view, the immune response does not begin with the recognition of foreign antigens, but by the reaction of auto-antibodies generated by the multiplication of their specific antibodies. The immune system reacts not to environmental antigens, but to what Jerne calls the "internal image" of these antigens, first because the immune system's antibodies already strongly express all possible antigens, like a mirror image of the antigen universe, and second because a systemic immune reaction is a reaction to some of the organism's antibodies (and not to antigens themselves) (Jerne 1974).

Jerne puts things even more radically a little over a decade later, upon receiving the Nobel Prize:

> I should therefore like to conclude that, in its dynamic state, our immune system is mainly self-centered, generating anti-idiotypic antibodies to its own antibodies, which constitute the overwhelming majority of antigens present in the body. The system also somehow maintains a precarious equilibrium with the other normal self-constituents of our body, while reacting vigorously to invasions into our body of foreign particles,

proteins, viruses, or bacteria, which incidentally disturb the dynamic harmony of the system.

(Jerne 1985)

These two texts (Jerne 1974, 1985) are quite useful, since they at once underline Jerne's main idea—that the immune system only ever reacts to its own disturbances—and show that Jerne does not set out to deny that the antigen's foreign nature is the decisive factor for immunogenicity. In other words, what Jerne intends to show is that any immune response presupposes an autoreactivity, and not that the discrimination between "self" and "foreign" is not the right criterion of immunogenicity (Jerne 1984). Jerne originates an idea that proved to be important for every critique of the self-nonself theory (i.e., the idea that autoreactivity is normal and necessary to the organism), but he himself does not consider his view a refutation of the self-nonself theory, and thus it is probably an exaggeration to call him the main critic of the self theory in the twentieth century, as does Alfred Tauber (Tauber 2000; see also the critiques of Anderson and Matzinger 2000c).

Thus, Jerne does call into question a central claim of the self-nonself theory, but without criticizing the criterion of immunogenicity it proposes. He refuses the proposition that the organism's own components (i.e., its "self") do not trigger immune reactions; instead, for him, self-reactions are frequent and even part of the immune system's normal functioning. In reality, Jerne does not at all deny that the foreign is immunogenic, but he considers that the foreign is only immunogenic inasmuch as it constitutes a disturbance of the immune system, which moves from an autoreactive state to another autoreactive state. In so doing, and notably through his concept of "internal image" (Jerne 1974: 383), Jerne simply reinforces the internalism already expressed by Burnet. I have, in effect, shown how Burnet transitioned from an open to a closed self; that is, the concept of an organism as the unity of a plurality interacting with its environment and capable of being influenced by it, to that of an organism as a homogeneous unit, coming from its genes and whose identity is defined early on and must subsequently be defended from any external threat. Now, Jerne is claiming that not only does the immune system defend the "self's" integrity, but that it is also only ever concerned with the "self," since it reacts to disturbances to the autoreactive equilibrium.

What should we make of Jerne's formulation? It has played a fundamental role in immunology from the 1970s up until today in that it gave rise to the idea of normal autoreactivity. From this perspective, it should not be underestimated. It is possible that without Jerne taking this position, few experimental researchers would have tried to find evidence of the organism's normal immune reactivity to its own components. At any rate, it is clear that the continuity theory takes from Jerne the need to account for normal autoreactivity and normal autoimmunity. The proposition Jerne advances in 1974 suffers, however, from an extreme vagueness. Jerne is the first to admit this: "The weakness of this developing network theory lies in its lack of precision" (Jerne 1974: 386). He believes that the efforts of future immunologists will be directed toward clarifying this theory; but if we are to judge by Jerne's direct heirs, whose theses I will shortly analyze, we might be tempted to think that the clarification has not, for the most part, been very complete. It is also important to note that although Jerne's general idea (that of the existence of a normal immune autoreactivity) was fruitful, the experimental foundations he proposes (and they are rather few) are not the ones that have been considered by the immunological community as establishing this idea. Indeed, the autoreactivity shown by immunologists in the 1980s involved T cells rather than antibodies (produced by plasma cells, which themselves originate from B cells), contrary to what Jerne had suggested.

Furthermore, with Jerne come the origins of a problematic tendency in systemic views of immunity: Jerne states very clearly that the immune system is exclusively, or almost exclusively, made up of lymphocytes: "the lymphocytes *are* the immune system, or at least 98% of it" (Jerne 1974: 377; Jerne 1976). This tendency, which, again, only reinforces one of Burnet's theses, is largely contradicted by contemporary immunology. As we have seen, recent results show that lymphocytes are inarguably important in vertebrates, but they are precisely only present in these organisms and, moreover, even in vertebrates the lymphocytes are but one of the actors of immunity—and, moreover, rarely used. Now, as soon as we propose that the activation of adaptive immunity depends on innate immunity, the essential question becomes: Is a normal autoreactivity possible in the components of innate immunity? This question, to which I have attempted to respond in the previous chapter, was never asked by Jerne, nor by the proponents of autopoiesis and self-organization, since all made the mistake of equating the immune system with the lymphocytic system.

A PROBLEMATIC CONNECTION BETWEEN THE IMMUNE SYSTEM AND THE NERVOUS SYSTEM

Jerne's second big idea is that the functioning of the immune system is very close to that of the nervous system (Jerne 1974: 387). The reasons for this connection are as follows: The two systems are very complex, integrating millions of stimuli every second; both can respond to these stimuli by an activation or an inhibition; and both are spread throughout the organism. Immediately, though, Jerne slips into a connection between immune system and cognitive system, which is not the same thing. This connection would be based on the fact that immune receptors are capable of "recognizing" some antigens, in the double sense of identifying these antigens (in distinguishing between their own patterns and those of others) and of remembering them for a more efficient response in case of a second encounter (this is the phenomenon of immune memory). Following Jerne, numerous connections between the immune system and the nervous system as a cognitive system were proposed, especially during the 1980s and 1990s. The suggestion was that immunology was going to move from a "defensive" paradigm to a "cognitive" one (Tauber 1997). Such a connection is found in Coutinho (e.g., Varela et al. 1988), as well as in Maturana and Varela (1980), and Irun Cohen (1992a, 1992b, 2000a).

From my perspective, this connection must be considered very cautiously. Immune interactions are biochemical, and we can certainly, for the sake of simplification, use cognitive terms to describe them—the most frequently used term by immunologists being that of "recognition"—but this linguistic facility certainly must not lead us to believe that any true cognitive act whatsoever occurs in immune reactions. Indeed, no immune "memory" exists in the strict sense of the word. The term is just a way of indicating that certain mechanisms ensure a more rapid and effective immune response in the instance of a second encounter with a same antigen. The use of these metaphors is not problematic as such, as long as there is a precise scientific term behind each one. Nevertheless, this connection between immunity and cognition has been considerably favorable to the problematic idea that the study of the immune system tells us something about our identities as conscious beings and even about our own personal identity (Howes 2000). Even though it is often difficult to do in practice, I suggest it is nevertheless important to avoid, when possible, using terms such as "recognition," "education," "memory," etc. in immunology.

THE IDEA OF AN IMMUNE "LANGUAGE"

Jerne's third idea is that immune interactions function as a "language." The concept of "information" is central to his work. In 1985, Jerne even suggested a connection between the functioning of the immune system and the concept of "generative grammar" proposed twenty years earlier by Noam Chomsky in the psycholinguistic field (Chomsky 1964). In both cases, says Jerne, it is a matter of an open-ended system (Jerne 1985: 1058). Here again, as far as it is only a metaphor used to underscore the huge potential for receptor production by the immune system, this connection between immunity and language is acceptable. But when this connection transforms into the idea that immunity itself is a language, as is the case with Cohen and Atlan who I will look at later on, then this connection seems to me harmful to the clarity of immunological concepts.

These ideas of Jerne's have had a considerable influence on modern immunology, at least until the early 1990s. The major evolution achieved between Jerne and his immediate successors is the movement toward the idea that the immune system *only* responds to the "self," and thus that the immunological "nonself" does not exist. Two principal trends come out of Jerne's immune network thesis: the first is that of autopoiesis, which is strongly internalist; the second is that of self-organization, which radicalizes the cognitive and linguistic aspects of Jerne's theory. I will now look closely at autopoiesis and self-organization in order to show that both reinforce the imprecision that was already present in Jerne's work and that they cannot be considered as immunological theories in the strict sense of the term.

Immunology and the Concept of Autopoiesis

One of the major offshoots of Jerne's ideas is the thesis of *autopoiesis*, first suggested by Maturana and Varela. Proponents of autopoiesis resorted to immunology to support their theses, but autopoiesis is a more general vision of all biology, bringing together immunological, neurological, and other questions (Maturana and Varela 1980). Autopoiesis was one of the most influential "systemic" biological theories in the 1970s and 1980s, even if it has been largely abandoned today.

Proponents of autopoiesis claim that the organism is "self-constructed" and "autonomous"—that it regulates itself its activity. Its adherents radicalized the closure of the "self" that was already showing up in Jerne. For Varela and his collaborators, the organism is always the product of its own creation, and nothing can influence it "from the outside": the organism modulates and interprets any external influence so that any interaction, including an immune one, is in fact an endogenous modification. There is no distinction made by the immune system between self and nonself because there is only the self. It may be disturbed either from the inside or from the outside (it does not matter), but nothing can affect the self without being taken into the complex network of endogenous immune interactions that precede it:

> According to the reasoning developed in the previous sections, all immune events are directed inward, not outward, and the organism perceives the penetration of foreign materials not by recognizing them as foreign, but rather because the foreign materials interfere with ongoing reactions which exist as links in a complex network of interactions. The organism responds to an "internal image" of the foreign molecule, to its meaning translated in terms of the language previously utilized by the network. Thus, in a way, all immune reactions are "autoimmune" (directed inward) and exogenous antigens are recognized by "cross-reactions."
>
> (Vaz and Varela 1978: 251–52)

Antonio Coutinho, who has explicitly claimed the role of Jerne's heir apparent, has pushed the theses of autopoiesis into a more experimental direction. Coutinho and his collaborators consider the immune system as connected with all of the organism's biochemical events, whose activity the immune system would regulate. They take up the central thesis of autopoiesis in opting for the term *autonomy* to describe this functioning (Coutinho et al. 1984).

The autopoiesis thesis, however, seems difficult to accept. First, it is not easy to assess to what extent the idea of autoreactivity is innovative. In effect, if the proponents of autopoiesis are content to reaffirm Jerne's thesis that all autoimmunity is not harmful, their assertion is true, but it does not bring anything new to the discussion. If, on the other hand, they radicalize Jerne's thesis by saying that there is only autoreactivity, that is, that the immune receptors *only* respond to the self (e.g., Coutinho et al. 1984: 164), this is an original but erroneous thesis. It is false to claim that the immune system can never be involved with exogenous antigens: to say that the immune system only responds to these exogenous antigens after having "processed" them is a truism, but to say that the immune system only ever encounters its own components is simply not true (think, for example, of a Toll-like receptor interacting with some lipopolysaccharide expressed by a bacterium). The thesis that there is only autoreactivity is thus false as soon as it is interpreted in its extreme form.

Second, the proponents of autopoiesis reinforce a difficulty that appeared in Jerne, that is, the exclusive attention to lymphocytes (e.g., Vaz and Varela 1978). Again, the question arises here of knowing whether or not the autopoiesis thesis can be applied to innate immunity components.

Third, in reality the autopoiesis thesis does not at all constitute an immunological theory in the sense of a structured and testable hypothesis. Its proponents begin by claiming that immunology must not be preoccupied with the detail of molecular and cellular mechanisms, but must on the contrary adopt a "holistic" perspective.

The difficulty, however, is that this holistic approach leads them to claim that the molecular differences between patterns expressed by antigens do not constitute the correct criterion for understanding what triggers an immune response. Yet at the same time, these immunologists do not offer any other criterion of immunogenicity. It is certainly not enough to say that the immune system responds to "disturbances of the system," since such a proposal is, at best, just a description of an activating immune response, rather than an attempt to give an explication of it. Proponents of autopoiesis are not offering a criterion of immunogenicity, they are not trying to say when and why an immune response is triggered, they are not trying to propose experiments that could invalidate or corroborate their hypotheses, and, as a result, they are not proposing, in the true sense of the word, a theory of immunity.

Thus, the thesis of autopoiesis uses both metaphorical and ill-defined terms while at the same time making an extreme case for the closed-off nature of the "self" already set forth by Jerne. Autopoiesis in no way offers a testable immunological theory, but rather a series of often insufficiently argued theses backed by a metaphysics of life and a general biological perspective that mixes explications of how thought functions and analysis of molecular processes as interactions between an antigen and an antibody. These problems, I think, illustrate why this thesis has fallen by the wayside today.

Autoimmunity, Self-Organization, and the Language of the Self

Irun Cohen represents another trend among systemic theories of immunology. Beginning in the 1980s, following in Jerne's footsteps, Cohen insists on the importance of autoimmunity in normal immunity. Joined at the end of the decade by Henri Atlan (Atlan and

Cohen 1989), Cohen suggests an interpretation of autoimmunity aided by the concept of self-organization (Atlan and Cohen 1998).

There are two very different ways of presenting Cohen and Atlan's work. The first is to consider them part of the group of scientists who, in the 1980s, drew the immunological community's attention to the importance of normal autoreactivity. From this perspective, their contribution, though valid, is nevertheless modest. The concepts had been proposed by Jerne long before and, although Cohen did perform several experiments that contributed to the demonstration of this autoreactivity phenomenon (e.g., Cohen and Werkele 1972; Moalem et al. 1999; see also Cohen 1992a for an overview), most of the experiments that clearly established this idea are not his doing. From this viewpoint, Cohen and Atlan's self-organizational perspective is useful, though not especially groundbreaking.

Another way to reflect on their work, however, is to be less interested in the general idea (normal autoreactivity), than in the specific theses that they support. I believe that only this second way allows for an evaluation of Cohen and Atlan's true contribution to contemporary immunology. From this point of view, their propositions are more original, but often imprecise or inexact; Cohen and Atlan are essentially proposing the following theses:

1. Autoimmunity is normal and essential to the organism's functioning.
2. The triggering of an immune response is caused by many factors, not one simple dichotomous mechanism (such as self/nonself).
3. The function of immune cells is not to reject nonself, but rather to maintain the body's functioning (Cohen 2000b).
4. Immune interactions are cognitive in nature and depend on the organism's immune self-knowledge, in particular on

the expression and knowledge of that which Cohen calls *homunculus* (Cohen 2000a), and which refers in fact to a limited group of dominant autoreactive antigens. Note that as Cohen himself recognizes, the existence of the homunculus is not at all proven, but it is "mainly a prescription of a research program" (Cohen 2000a: 204).

5. Immunity is a language (Atlan and Cohen 1998).

These five theses seem problematic to me. In the first, Cohen and Atlan confuse autoimmunity and autoreactivity: they are suggesting that lymphocytes constantly respond to the "self" in the form of genuine immune responses, but that these responses are inhibited by an entire series of mechanisms that prevent the organism from self-destruction. Immunology in the 1990s demonstrated otherwise: It established that lymphocytes are autoreactive, that is, that they are in continuous, weak, interactions with the "self's" antigens. There is, for these lymphocytes, autoreactivity, but not autoimmunity in the sense that an effector immune response would be triggered and subsequently be inhibited. Thus, the demonstration of normal autoreactivity in the 1990s did not at all confirm Cohen and Atlan's idea of a normal lymphocytic autoimmunity. Nevertheless, if immunologists demonstrate in the near future that autoimmune lymphocytic immune responses are frequent but inhibited by regulatory responses such as those that are due to regulatory T cells, then Cohen and Atlan will have been proved correct. But in the current state of affairs, we cannot say that this is the case—unless we wish to confuse autoreactivity with autoimmunity.

The second and third theses do not merit criticism for false premises, but rather for their imprecision. To refuse "simplistic" and "dichotomous" explanations is easy and leaves little room for error. Nothing, however, is further from a scientific theory than such declarations of the immune system's extreme complexity. Cohen often insists on the multifaceted nature of an immune response (e.g., Cohen 2000b). Faced with such complexity, we can only conclude that the triggering of an immune response is scientifically unpredictable. In other words, Cohen and Atlan are also abandoning any search for a criterion of immunogenicity. They are not offering a theory of immunity, but a "way of looking" (Cohen 2000a: xvii) at immune phenomena that insists on their complexity, reflexivity, and self-organization. I agree largely with Anderson and Matzinger, who write in a critical evaluation of Cohen's thesis: "because Cohen lays out no rules, makes no predictions, suggests no cellular interactions, and says only that the immune system must somehow judge the context in which it sees an antigen, we cannot properly analyze or comment on his model. It isn't a model. It isn't even a map [of the immune system]" (Anderson and Matzinger 2000b: 289). The authors then conclude: "It may be more comfortable to luxuriate in boundless complexity, and it is certainly safer, but it is less likely to spark clear experiments."

As far as the fourth and fifth theses are concerned, the usefulness of cognitive and linguistic metaphors never clearly appears. How exactly do these anthropomorphic and teleological comparisons help us understand the immune system's functioning? If they constitute a way of recalling the existence of autoreactivity and immune "memory," then they are useless, since these two phenomena are already well known. If they constitute the foundation of a strong informational vision of life, then they must be rejected, since such a vision may lead to erroneous ideas, as has been demonstrated in other contexts (Griffiths 2001). In their 1998 article, Atlan and Cohen discuss the "cognitive creation of meaning" of the immune system and write: "The particular response we observe is the outcome of an internal process of weighing and integrating information about the antigen.... A cognitive immune system organizes the information borne by the antigen stimulus within a given context and creates a format suitable for internal processing; the antigen and its context are transcribed internally into the 'chemical language' of the immune system" (Atlan and Cohen 1998: 714). So, if one takes the theses advanced by Cohen and Atlan seriously, they offer a rather imprecise and often invalid vision of the immune system's functioning.

Systemic Thesis and the Continuity Theory

From the perspective of the demonstration of autoreactivity, systemic theories have played a useful role, and my continuity theory draws much of its inspiration from them. The general thesis that the immune system only functions if it is constantly stimulated by endogenous components has revealed itself to be quite accurate. Still, it is necessary to make two observations about this thesis. First, the proponents of systemic theories have offered an idea rather than a demonstration, following Jerne in the notion that the immune system could be seen as an autoreactive network; however, very few of these systemic theorists offered major experimental demonstrations of autoreactivity. In addition, they were often imprecise with their terms, particularly with the issue of autoimmunity itself, which was not sufficiently distinguished from autoreactivity.

Furthermore, by only criticizing one of the two central claims of the self theory (i.e., "the organism does not trigger an immune response against its own components"), systemic immunologists reinforced the second (i.e., "the organism triggers an immune response of rejection against any foreign entity") and in so doing exacerbated the privileging of the "self." Indeed, by claiming that there was only the self, they reinforced the internalist characteristic inherent to the self theory. The continuity theory's objective vis-à-vis these systemic views is to *open up* the immune system to its environment instead of viewing it as exclusively self-centered. The immune system cannot be considered as closed and self-centered precisely because of the illustration, in the course of the 1990s, of tolerance phenomena that I have relied on, illustrations that I consider to have had the most revolutionary impact on contemporary immunology. In contrast, these same phenomena have elicited no interest from systemic immunologists, especially those studying self-organization.

Finally, and more fundamentally, the systemic concepts do not offer any criterion of immunogenicity. They remain on a descriptive, even interpretive, level, but they do not explain the reasons why an immune response is triggered.

COMPARISON WITH THE "DANGER THEORY"

Starting in the 1990s, and under the influence of her friend Ephraim Fuchs, the American immunologist Polly Matzinger proposed a concept of immunity designed to replace the self-nonself theory that she called the "danger model" (or "danger theory") (Matzinger 1994). This theory met with an enthusiastic response and great success, giving rise to publications in the most renowned international journals (e.g., Gallucci, Lolkema, and Matzinger 1999; Matzinger 2002). Although the enthusiasm surrounding Matzinger's ideas has lessened today, I think it is indispensable to situate the continuity theory in relation to the danger theory, since it relies on new concepts and experiments that make up an important attempt at critiquing the self-nonself theory.

The main proposal in the danger theory is that every immune response is due not to the presence of "nonself," but to the emission, in the organism, of "danger signals." I will present the main points of Matzinger's proposal before moving on to an explanation of why I think that her theory is ultimately unsatisfactory due to its lack of a clear definition of "danger." I suggest that the danger theory claims to resolve the problem of a criterion of immunogenicity with the help of a key term ("danger") that, in reality, explains nothing and in fact demands an explanation of its own.

The Principles of the Danger Theory

In her founding article on the danger theory, Matzinger (1994) claims that if one moved from a vision of immunity based on self-nonself differentiation to one where every immune response is actually due to the emission of danger signals, then most of the mysteries of modern immunology would be solved. The immune system, according to her, reacts to a group of danger signals sent by damaged tissues or cells, most of the time in response to *exogenous* agents, but which could also be emitted in response to *endogenous* agents (cellular stress, autografts, etc.). For example, fetomaternal tolerance would be explained by the fact that the fetus is not "dangerous" for the mother, and the same would go for the tolerance of commensal or symbiotic bacteria, normal autoimmunity, etc.

The danger theory's starting point is to ask what triggers the activation of B and T cells. At the core of all of Matzinger's theoretical writings lies the "two signal" theory (Bretscher and Cohn 1970; Lafferty and Cunningham 1975). Matzinger relies on this theory to suggest the following thesis: since any immune response starts with the activation of an antigen-presenting cell (APC), we must understand the mechanisms of this activation. According to Matzinger, the large number of exceptions to the rule of self-nonself differentiation shows that APCs do not recognize "nonself"—instead, they recognize "danger." Her theory is best understood from the perspective of a double opposition (Matzinger 2002): to the classic self-nonself theory, on one side, and to the *infectious nonself* theory, on the other. This second theory was proposed in 1989 by the immunologist Charles Janeway as an extension, not a critique, of the self-nonself theory (Janeway 1989, 1992). According to

Janeway, APCs have their own form of distinction between self and nonself and can recognize evolutionarily conserved pathogens. These cells would have evolved to interact with microbial patterns widespread in nature, such as lipopolysaccharide. Antigen-presenting cells would not recognize any "nonself"; instead, they would only recognize these foreign patterns that are highly conserved throughout evolution. Thus, Janeway proposes that the immune system distinguishes between "infectious nonself" and "non-infectious self." Matzinger's theory builds upon Janeway's but also tries to go beyond it by completely getting rid of the self-nonself vocabulary. The claims of the danger theory are the following: first, the distinction between self and nonself is not relevant to immunology; second, this self-nonself distinction should be replaced with that of benign versus dangerous (emission of alarm signals or "*danger* signals" by damaged tissues or cells); third, these "danger signals" help explain those phenomena that the self theory cannot, such as tolerance of the fetus, autoimmunity, differences in transplant acceptance, etc.

The Advances Introduced by the Danger Theory

The danger theory introduces important advances on the theories that preceded it. It offers arguments and experimental results to argue that the self-nonself distinction is not the proper criterion of immunogenicity. Building on Janeway's thesis, the danger theory put APCs at the center of the triggering of an immune response. Lastly and above all, from a scientific perspective, Matzinger and her colleagues' most successful texts lay out the considerable advantage of looking for a criterion of immunogenicity (contrary to the systemic theories), and propose experiments that could settle the disagreement between competing theories (see, in particular, Anderson and Matzinger 2000b, 2000c).

Criticism of the Danger Theory

THE ABSENCE OF A DEFINITION OF "DANGER"

In its basic form, the one that raised as much enthusiasm as criticism, the danger theory claims that every immune response is triggered by the emission of danger signals by immune cells. But how is "danger" defined? Matzinger and her colleagues never precisely answer this question. Instead, they prefer to use several apparently equivalent terms: "danger," "damage," "stress," "necrosis," "inappropriate cell death," etc. Yet these terms are not synonymous since, for instance, a cell can die from necrosis without causing damage to the organism's tissues. Therefore, the accumulation of these heterogeneous terms appears as a way of accounting a posteriori for the triggering of any immune response. It would appear that Matzinger and her collaborators wanted to propose an idea (all "danger" is immunogenic), while keeping for later the task of defining what "danger" precisely means. For instance, at a time when much work was being done on heat-shock proteins (HSP), Matzinger suggested that these proteins could very well be the "danger signals" she had described (Matzinger 1998). Yet, to be testable, the danger theory's propositions must be much more precise about what counts as a "danger signal."

In addition to being too vague, the term "danger" is anthropomorphic at its core, which is problematic for a theory that aims specifically to criticize this very aspect of the self-nonself theory. How could immune cells perceive "danger" as such (Silverstein and Rose 1997)? The "danger" is likely to be a projection on the immune system of the immunologist's perception. It is easy to allow, for instance, that a pathogenic bacterium or virus is "dangerous" (this is practically the definition of a pathogen) and that some commensal bacteria are not dangerous (this is what "commensal," "*eating together*," means or, rather, implies), but how to understand, among other things, immune responses to transplants? An organ transplant is useful for the receiver, but Matzinger says that it is "dangerous" because the surgeon's gesture damages the patient. Furthermore, how does the notion of "danger" explain the immune system's responses to innocuous antigens such as allergens or food antigens? One answer could be that the immune system "sees" these as dangerous, but this exposes the circular logic of the thesis.

Thus, the difficulty stems from the imprecision of the term "danger": it cannot establish a theory of immunity, since it does not offer a solid criterion of immunogenicity. There is, however, another possible conception of the danger theory that allows for a much more convincing explanation of the immune system's functioning.

DAMAGES AND IMMUNE RESPONSE

Although Polly Matzinger has always held to the "danger" rhetoric, her theory is much more relevant if interpreted as follows: any immune response is due to *damages* to the organism's cells or tissues. This is actually the interpretation that Matzinger herself proposes when she describes the molecular details of her theory (Matzinger 1994; Anderson and Matzinger 2000a; Matzinger 2002). Indeed, if it is presented this way, Matzinger's theory is certainly less original but much more solid. Less original because it constitutes in large part a return to older ideas of Metchnikoff (e.g., Metchnikoff 1892) and Ehrlich (e.g., Ehrlich 1897; for a critique of Matzinger on these grounds, see Silverstein 1996). Alfred Tauber wrote on many occasions that he agreed with Matzinger's theory, most notably because of this convergence between her work and Metchnikoff's on the central role given to the tissue damage in the organism to explain the immune response (see, e.g., Tauber 2000). In addition, Matzinger and her colleagues admit that their theses comprise a way of "breathing new life" into some of Ehrlich's ideas that had been ignored because of Burnet's dominant self-nonself thesis (Fuchs, Ridge, and Matzinger 1996). This more subdued originality in Matzinger and her collaborators' proposals contrasts with some of their claims for a radical paradigm shift in the way of the Copernican revolution, proclamations that explain in large part the publication of numerous articles loudly criticizing the danger theory, often with a somewhat exasperated tone (e.g., Janeway et al. 1996; Silverstein and Rose 1997; Vance 2000).

Yet the question of originality is ultimately of little importance to me. What is at stake is to evaluate the validity of the statement that "any immune response is due to damages inflicted on the organism's tissues" as such. This thesis is much more precise and relevant than the preceding one ("any immune response is due to the emission of danger signals in the organism"), since inflammation, and more generally damages to the organism, accompany a large number of effector immune responses. Moreover, it is important to emphasize that according to Matzinger these damage-based immune responses occur at the level of tissues, because in recent publications she has insisted that immunity should be thought of at the level of tissues rather than at the level of cells (Matzinger 2007; Matzinger and Kamala 2011).

When expressed this way, Matzinger's theory seems much clearer to me. Yet I think it is still erroneous, for the following reasons:

1. *The imprecise definition of damage signals.* Despite the clarity of the general thesis, what counts as a "damage signal" from a molecular perspective is not satisfactorily defined by proponents of the danger theory. For example, if damage is the criterion of immunogenicity, it is impossible, though Matzinger does, to move without any transition to the idea that all necrotic cell death is immunogenic. Matzinger never explains how immune cells could "perceive" damages inflicted on the organism's tissues. This brings me back to

the concern I have regarding the "danger signals": what exactly are "damage signals" from a molecular point of view? Nonetheless, a possibility could be to accept the thesis as such while waiting, as do Matzinger and her collaborators, a precise experimental determination of these "damage signals."

2. *The possibility of immune response without damage.* Certain antigens seem to provoke an immune response without any damage preceding it; this is the case with the interaction between immune receptors and pathogen-associated molecular patterns or PAMPs (Medzhitov and Janeway 1997; Vance 2000), which plays a major role in innate immunity (which is itself central, or should be, in Matzinger's work in that it triggers the adaptive response). Damages do not explain the activation of macrophages and dendritic cells, the first immune cells to respond to antigens. Contrary to one of Matzinger's central claims, pro-inflammatory signals are not enough to activate dendritic cells (Spörri and Reis e Sousa 2005; Joffre et al. 2009)—this activation requires the specific interaction with an antigen. Moreover, several experiments suggest that grafts achieved without any inflammation or damage still can trigger a strong immune response of rejection, which underscores that damage alone is not a valid criterion of immunogenicity (Bingaman et al. 2000). Finally, it seems very difficult to explain the activation of regulatory immune mechanisms such as regulatory T cells by the suggestion that they would interact with damage signals (Matzinger's collaborator Colin Anderson himself seems to favor a quantitative explanation: Anderson 2009).

3. *Most of the time the damage originates in the immune system.* Damages to the organism's tissues are most often due to immune cells themselves. They are typically provoked by macrophages, which release pro-inflammatory cytokines, and then by the activated lymphocytes, which always provoke some damage to the organism, and which, in the absence of regulatory mechanisms like

regulatory T cells, would even cause major damage. For instance, the T *helper* 17 cells, which produce interleukin 17, are major mediators of inflammation (Bettelli et al. 2007). In addition, it was recently shown that certain immune mechanisms (activation of the complement via M immunoglobulins) are responsible for the aggravation of damages to the organism caused by burns (Suber et al. 2007). The immune system is therefore more a *source* of damage than a detector and suppressor of damage. Two major problems arise here. First, if any immune response is caused by damage and any immune response causes damage, then the organism should enter into a vicious circle of immune activation, almost regardless of an antigen's presence. Second, the organism's first immune cells to react to an antigen (mainly macrophages) release the first pro-inflammatory signals, so the question for the immunologist is: What activated in the first place these cells that are themselves the root of the damage? It cannot be the damage itself. In the continuity theory, this activation of the first cells is explained by the molecular discontinuity. The danger theory, however, confuses an effect of the immune response with its cause: inflammation and tissue damage in the organism do indeed accompany an immune response, but they do not trigger it.

4. *The impossibility of firmly replacing "nonself" with "damage."* Matzinger is clearly mistaken when she systematically replaces the nonself argument with the damage argument. For example, she claims that immune responses of rejection to grafts are not due to the recognition of an antigenic difference, but rather to damage incurred by the surgery needed to perform the transplant (Matzinger 1998). If this were true, how is it that a surgical autograft does not trigger a rejection? Furthermore, how is it that transplant rejection occurs in nature (without surgical intervention), as in the case of rejection reactions between two colonies of *Botryllus schlosseri* (De Tomaso et al. 2005)?

5. *Some errors in major experimental data.* Matzinger and her colleagues misinterpret several important experimental data. Their largest error involves cancers. They claim that the danger theory proves its superiority to other theories most when it comes to the study of tumors: only the danger theory, according to them, explains why there is not an immune response to tumors. To begin with, this claim is incorrect from a historical point of view: the self theory did begin by claiming that it was normal to not have any immune response to tumors because they constituted part of the "self" in that they come from the organism's genome (Burnet and Fenner 1949), before going on to claim that tumoral patterns were part of the "modified self" that triggered an immune response. Second, and even more important, the immune system actually does respond quite well to developing tumors, as we have seen in the previous chapters. Matzinger and her collaborators relied on old experiments, which tried to show that immunodeficient mice would not develop tumors any more frequently than normal mice; these results have since been shown to be inaccurate (Dunn et al. 2002; Dunn, Koebel, and Schreiber 2006; Koebel et al. 2007; Bindea et al. 2010). Numerous recent data show that the immune system responds to tumors and indeed eliminates a large number of them (or prevents their development). Of course, supporters of the danger theory could say that the immune response to tumors is explained by the fact that tumors can cause "damage" to the organism, but this would be a total contradiction to what they have maintained until today and would again show the extreme flexibility of their concept of "damage."

I conclude then that the term "danger" is not precisely defined. At its best, it is a synonym for "damage" (caused to the organism's tissues), which is experimentally inadequate. This is probably why Matzinger explicitly avoids equating "danger" with "damage" and prefers instead to use several expressions as synonyms even though they are not ("danger," "stress," "damage," "abnormal," etc.) This makes possible to justify a posteriori any immune response, but it does not contribute an exact response to the question of a criterion of immunogenicity—although Matzinger and her colleagues initially frame the question quite well. Like Matzinger and colleagues, I think that the most important quality of an immunological theory is its conceptual clarity and its ability to be tested experimentally (Anderson and Matzinger 2000b); as a result, I must reject the uncertainty of "danger," "stress," etc. Moreover, examining Matzinger's clearest proposition (that every immune response is triggered by damages to the organism), I believe I have shown that this view does not resolve appropriately the problem of finding a criterion of immunogenicity.

Is the Danger Theory an Extension of the Self Theory?

The danger theory repeatedly presents itself as a radical paradigm shift with regard to the self theory (Fuchs, Ridge, and Matzinger 1996). At the same time, however, Matzinger insists in several articles on the way her theory builds upon the self theory, improving it without claiming to change it from top to bottom (Matzinger 2002). She offers a history of self theories, the last two chapters of which are, first, Janeway's "infectious nonself," and, second, the danger theory itself. Obviously, a "renewed" sense of self (Matzinger 2002) is quite different from a radical revolution upending existing hypotheses. In accordance with an attitude often observed in the history of science, the danger theory constantly wavers between the proclamation of its extreme novelty and its connection to the long line of theories that preceded it. Is it possible to decide between these two tendencies? Is the danger theory a genuine rupture with the self theory, or is it simply a logical extension of it? In what follows, I suggest a balanced answer to this question. On the one hand, the danger theory is a major, though flawed, break with one crucial point of earlier immunology, that is, the importance given to the question of biochemical specificity. On the other hand, it builds upon the self theory on one essential point: the central problem remains understanding how the organism learns to tolerate its own components.

On one specific point, the danger theory is the furthest view possible from the self theory: the danger theory completely ignores the question of the antigen's molecular structure. In particular, it sets aside the question of the precise biochemical conditions (specificity, avidity, affinity, etc.) of a given immune interaction. The only thing that counts is the emission of "danger signals." Molecular difference (or resemblance) has no importance, whereas this is the central issue for both the self and continuity theories. This affirmation is even what actually seems to lead Matzinger to reject self-nonself differentiation: because a damage can perfectly be endogenous, the "foreignness" of an antigen does not cause the immune response. The idea is that if a tissue is severely damaged, even in the absence of a biochemical interaction with strong specificity, affinity, and/or avidity, there will be an effector immune response. Biochemistry thus becomes a negligible scientific field for understanding immune reactions. This, I think, would be a serious mistake. If the question of specificity, affinity, and avidity is set aside, how can we explain why some transplants are better tolerated than others? Why do so many pathogens escape the immune response using molecular mimicry? These data and many others illustrate the importance of precise biochemical research on immune interactions—which is exactly what has taken place for more than a century in immunology (Morange 1998, 2005).

I do think, however, that the danger theory is an extension of the self theory in the central issue that it raises of understanding how the organism learns to tolerate its own components. The danger theory effectively focuses on the adaptive immunity of lymphocytes; and because it fails to provide a sufficient explanation of the activation of cells that first react with an antigen, it does not offer a satisfactory explanation for the existence of tolerance to exogenous entities.

Matzinger focuses her attention on lymphocytes' activation. The central question in all of her articles is that of the "second signal" that lymphocytes must receive to be activated. Her answer is that this second signal is caused by the action of dendritic cells, which are activated when the organism is damaged. In concentrating this way on the "second signal," Matzinger extends the self theory's line of inquiry and commits two errors. The first is to underestimate the significance of discoveries on innate immunity—which not only forced immunologists to rethink lymphocyte activation (although this is an important point) but also constituted a reflection on the extension of immunity mechanisms (i.e., many more organisms than was previously believed have immunity). In a telling article, Matzinger explains her surprise when Charles Janeway asked her to compare contemporary reflections on innate immunity to her danger theory. Her remarks clearly show that in her mind the danger theory takes adaptive immunity as its sole subject: "How would one compare the system of cells and molecules that make up the body's first line of defense against pathogens with a model that attempts to lay out the adaptive immune system's guidelines for immunity and tolerance?" (Matzinger 1998: 399)

Later on, she insists that the link between the two is that innate immunity plays a decisive role in the triggering of adaptive immunity; however, this only confirms that the danger theory is, in the eyes of its creators, only interested in the activation of lymphocytes.

The second error is to stop short on the causal chain: it is vital to understand why the APCs are activated and, as I have shown earlier, to claim that it is because of damage inflicted on the organism is insufficient, since in many cases the first immune cells to respond to an antigen (macrophages and dendritic cells, in particular) are activated in the absence of damage and are themselves a source of pro-inflammatory signals.

In addition, the danger theory talks often of immune "tolerance," but each time, as with the self theory, it is actually referring to self-tolerance, that is, tolerance to the organism's endogenous components: "The Danger model is based on the idea that the ultimate controlling signals are endogenous, not exogenous. They are the alarm signals that emanate from stressed or injured tissues" (Matzinger 1998: 400). The danger theory extends the endogenous logic suggested by the systemic theories, as its preoccupation is explaining why the recognition of an endogenous antigen by a lymphocyte at the periphery generally leads to self-tolerance and not to an autoimmune response. It is thus a simple extension of Burnet's view to the question of peripheral tolerance. The difference with respect to other conceptualizations is that Matzinger claims that "self" antigens do not trigger an immune response because they do not cause damage to the organism (and not by virtue of their endogenicity alone), but the reduction of immune tolerance to self-tolerance is the same in both cases. Indeed, the danger theory is not truly a theory of immune tolerance, one that would consider tolerance to exogenous entities. Because the theory stops short of an explanation, the criterion of danger does not account for the discrimination between tolerated exogenous entities and those that are rejected: for example, it may seem appealing to say that commensal bacteria are tolerated (in contrast to pathogenic bacteria) because they do not cause damage to the organism, but in reality most pathogenic bacteria are recognized and rejected by the innate immune system before they have a chance to cause any damage. The criterion of damage is ultimately not appropriate to understanding the very first stage of an immune response. Consequently, the danger theory makes a mistake with regard to the first revolution in contemporary immunology—the decisive role of innate immunity—and it ignores the second revolution of understanding tolerance mechanisms to exogenous antigens. My conviction is that the continuity theory is the only immunological theory that offers an explanation for immune tolerance to exogenous antigens.

Thus, despite its considerable interest and the important clarifications that it has brought, the danger theory distances itself from the self-nonself theory on the wrong points (especially biochemical specificity), while extending it on the points that most deserve to be criticized (endogenicity, reduction of tolerance to self-tolerance). For all these reasons, I consider the danger theory inadequate as a theory of immunity.

In conclusion, among the self-nonself theory, the systemic theories, and the danger theory, none offer a truly satisfactory explanation of how the immune system works. The continuity theory that I offer relies on those previous theories, but it is also an attempt to assert a different, more exact explanation of immunity.

At this stage, another problem arises. The continuity theory attempts to find a precise criterion of immunogenicity while refusing the self-nonself criterion. It is not clear, however, whether it is possible to reject the self-nonself theory and at the same time keep the idea that immunology has something useful to say about the notion of biological identity. Does rejecting the self and the nonself require the rejection of identity as a concept? Does the continuity theory allow for a new discussion of the notion of biological identity? Or should we abandon the idea that immunology has anything to say about identity? If we still want to assert that there is a link between immunity and identity, what exact meaning should this word take on? In the final chapter of this book, I attempt to answer these questions.

Chapter 6

What Is an Organism?
Immunity and the Individuality
of the Organism

Contemporary immunology has been elaborated on the conviction that it could clarify the concept of identity as applied to living
things. The philosophical problem guiding this book is to evaluate the validity of this proposition by attempting to determine the
possible meanings of the term "identity" as it is used in biology.
Yet the demonstrations that I have made in chapters 3, 4, and 5
establish that the immunological self-nonself theory is inadequate
and should instead be replaced by the continuity theory; even the
self-nonself vocabulary sows too many seeds of imprecision to be
kept. As a result, my approach could give the impression that it
rejects immunology's ambition to articulate a scientific discourse
on biological identity. If immunology must effectively get rid of the
"self," then can it still claim to study the identity of living things?
This chapter responds affirmatively to this question. I believe that
the self is just one way—and an inadequate and confusing one,
at that—to discuss biological identity. However, this chapter's
fundamental goal, which also ultimately supports the thesis that
I have just set forth, is to clearly and meaningfully lay out the exact
notion of identity that immunology can help define. As I have

insisted from this book's first pages, contemporary immunology sets out to clarify the identity of living things, but without ever precisely defining what meaning it gives to this term. In this chapter, I aim to show that:

1. Immunology sheds light on only one meaning of the notion of identity as applied to living things, that is, identity understood as the organism's *individuality*. I will try to demonstrate that, though limited, this contribution is critical, as immunology brings a decisive answer to the question "What is an organism?"

2. With regard to other definitions of the term "identity," immunology does not bring a new or decisive contribution. This part of the argument concerns, in particular, identity understood as uniqueness and, more generally, as the description of individual characteristics. From this perspective, the notions of "self" and "nonself" and even at times "identity" as immunologists have used them since Burnet are much too equivocal, since they tend to suggest that immunology can contribute something important to the understanding of all aspects of biological identity, which is not the case.

I am thus coming to the ontological aspect of my work—perhaps the most important one from a philosophical point of view. Despite a long philosophical tradition that demands a total separation between ontology or metaphysics and experimental science, the link between the two seems to me close in two ways. First, in their work, scientists leverage metaphysical conceptions and concepts. The discussion of "self" threaded throughout this book is one example of a metaphysical concept that plays an important role in a scientific field: immunologists borrowed it from psychologists, who first borrowed it from philosophers. Moreover, these meta-physical concepts as they are used by scientists convey second-ary implicit ideas, often without awareness of the presuppositions involved in a term's use: for example, the adoption of the word "self" in immunology has contributed to an "insular" vision of the organism, since it is conceived of as self-constructed and closed to its surroundings (Tauber 1994; Pradeu and Carosella 2006a).

Second, experimental sciences construct a metaphysics and can thus be useful to metaphysicians' reflections. To be sure, if we define metaphysics as that which is beyond experience, then the experi-mental sciences would not be able to help clarify our metaphysics. But if metaphysics is a general description of the world, if it is "the science that deals with the most general, fundamental, and ultimate aspects of reality" (Ghiselin 1987), then the experimental sciences may offer an important perspective on metaphysics (Quine 1951, 1969; Ladyman and Ross 2007). By highlighting the link between metaphysics and experimental sciences, I am following a trend in contemporary philosophy, created as a response to the excess of criticism of metaphysics by logical empiricism. Here, "metaphysics" is therefore understood as a synonym for "ontology"—the task of metaphysics being to describe the "furniture of the world."

Several philosophers of biology (e.g., Hull 1989; Gayon 1996; Okasha 2006; Godfrey-Smith 2009) and biologists (e.g., Ghiselin 1987; Mayr 1987; Maynard-Smith and Szathmary 1995; Michod 1999; Santelices 1999; Queller 2000; Gould 2002; Gardner and Grafen 2009; Queller and Strassmann 2009; Folse and Roughgarden 2010) have analyzed the metaphysics that can be inferred from the theory of evolution by natural selection, in particular by show-ing which definition of the individual the theory of evolution implicitly bears. Attempts to show what physiology and molecular biology can bring to metaphysical concepts and problems is much more rare; this, however, is what I would like to do here, by showing to what extent immunology clarifies the metaphysical notions of identity, uniqueness, and individuality.

In this chapter, I will illustrate the metaphysical questions that immunology can and cannot answer, or cannot answer specifically. I will then move on to how immunology can contribute a precise and new response to the question "What is an organism?" Before that, however, we need first to address the following question: What does it mean to ask what a living thing's identity is?

IDENTITY, UNIQUENESS, INDIVIDUALITY: WHAT QUESTIONS DOES IMMUNOLOGY ANSWER?

In its most general form, the question of identity as it is applied to living things is: "*What* is a living thing *X*?" This question may be understood in two distinct ways that correspond to differing understandings of the notion of identity:

1. The question of a particular living thing's *individual description*, or "What is *this* living thing?" I will discuss later on how this question includes that of what makes up the individual organism's *uniqueness*.
2. The question of biological individuality, or "What counts as *one* living thing?" which is similar to the question "What counts as one *individual* in the living world?" Individuality is not the same thing as uniqueness, or even more generally as the description of the individual.

In what follows, I will examine these questions, keeping in mind the goal of determining whether immunology truly has any light to shed on them. Before I begin, though, it is important to underscore, in keeping with my objective as it was laid out in the introduction to this book, that I am using the expression "living thing" in its most general way. A living thing is any living entity, no matter what it may be. A living thing is thus not necessarily an organism. There are many examples of this. A cell that belongs to a multicellular organism is a living thing without being an organism. A species, a taxon, a population, a group, etc. can also be considered living things without, of course, being organisms. A living thing is simply an entity that lives.

To begin, then, we can understand the question "What is a living thing?" as dealing with an *individual description*, equivalent to "What is *this* particular living thing?" The question thus becomes that of what makes up the identity of a specific living thing; that is, what the characteristics of this living thing are. I think that the principal mistake of contemporary immunology, caused by the imprecision of the terms it uses, and especially of the word "self," is to think that it answers the question of identity *qua* individual description.

To ask what constitutes the identity of a particular living thing amounts to raising the question of what its own proper characteristics are—both those that it shares with others and those that make it unique. It is an extremely broad question, with a synchronic, as well as a diachronic, component (many traits, whether they are unique to one living thing or general, are acquired during development or during the organism's lifetime). This question may also be applied to different levels (morphological, cellular, molecular...) and can reach ever more precise levels of detail.

Even if one admits that biology can raise such an expansive question, it should be clear that no one branch alone can answer it. Some have suggested that immunology was precisely the biological discipline that could answer the question of identity as individual description. Those who make this recommendation are often thinking of the field of transplantation which, as we have seen, played a crucial role in the emergence of the idea that immunology defines the organism's identity and its individual difference. Leo Loeb writes, for instance: "We believe, then, the conclusion is justified that in certain respects these chemical differentials of organisms are the most characteristic features of individuals as such, and that in their totality and interaction they constitute the most essential biological basis of individuality" (Loeb 1937: 5). Many immunologists, moreover, have spoken of a complete immunological "self-knowledge," assuming that, in order to be able to distinguish "self" from "nonself," the immune system would have to know all of the organism's individual characteristics: "The individual, through the mediation of its immune system, seems to have a complete, encyclopedic knowledge of that which comprises it" (Claverie 1990: 35–36).[1] Jean Dausset (1990), as we have seen, defined the HLA system as the organism's "identity card." Yet as soon as one understands that the question of individual description involves both the individual organism's shared and unique traits, it quickly becomes clear that every (or almost every) biological discipline has something to say about it. I am thinking of several examples, such as morphology and comparative anatomy. Furthermore, any approach to biological classification sheds light on this question in its own way. Generally speaking, developmental biology, genetics, and almost all subfields of biology, can, to different degrees, contribute something important to this issue of individual description. Immunology's contribution may be real, but it is certainly not exclusive or even the most significant.

There is, however, a way of phrasing this question that could assign a privileged role to immunology. Identity as an individual description includes, but is not reduced to, identity as *uniqueness*. If, now, the question of a living thing's identity is limited to the question of its uniqueness, doesn't immunology then provide a particularly useful clarification?

Immunology effectively demonstrates the individual organism's uniqueness. It even reflects one of the main arguments used in the introduction of this book in order to raise the problem of biological identity as it appears in immunology. Certain immunologists see in the explication of each organism's uniqueness the main goal of their discipline (Medawar 1957; Cohen 2000a). Immunology shows, in effect, that even two genetically identical individuals are different from the point of view of their immune system. The reason for this is that certain organisms build their immune system throughout their entire life: this is the case in all organisms that have "immune memory," that is, the ability to trigger a more effective and rapid immune response when encountering an antigen for the second time. These organisms develop specific immune receptors for the antigen during their first interaction with it. Nevertheless, there is an important reservation that should be raised at this point: all organisms do not have this quicker and more effective response upon the second encounter. This could have dealt a fatal blow to the case for immunological uniqueness a few years ago, since it was thought that only higher vertebrates possessed this mechanism. Yet as I have shown (chapter 1), immunologists recently found evidence of immune memory in a large number of organisms, so much so that immunology can explain the uniqueness of numerous organisms, including genetically identical ones (which is the case in asexually reproducing organisms, including some plants).

Consequently, immunology does have important contributions to make to the question of a living thing's identity understood as its uniqueness. However, a second, more conclusive reservation comes into play: in reality, it is development as a whole that is involved in the construction of the living thing's uniqueness over time. By this I mean development in the largest sense of the word as all of the processes that build the living thing over time, rather than the narrower sense of the living thing growing from an immature state to a mature one. As Gilbert (2010: 541) put it, "development never ceases." These are in fact very general processes, such as developmental epigenetics and phenotypic plasticity (West-Eberhard 2003), that account for the ways in which a living thing constructs its uniqueness over time: it is only as the study of one example of these developmental processes that immunology contributes to the understanding of a living thing's uniqueness. The nervous system in a vertebrate organism is a particularly significant example: the degree of uniqueness which, in a vertebrate, is due to the particular neural connections that it establishes in the course of its existence (in response to stimuli from the environment) is higher than that due to particular immune receptors that it acquires.

Immunology thus addresses the uniqueness of each living thing quite well; however, it is certainly not the only scientific branch to do so, since all of the biological disciplines that give individual development an important role also tackle this question. Immunology is also by no means the biological field that demonstrates the living thing's uniqueness with the most convincing arguments.

It is also important to underscore that immunology's approach to understanding a living thing's uniqueness has nothing to do with the self-nonself theory. This uniqueness is experimental data that all immunological theories must account for, but which does not exclusively belong to a single theory and is not better explained by one theory than by any other. In other words, the "self" of the self-nonself theory does not refer to the living thing's uniqueness and has nothing to say in particular about it: the immune self is the group sum of components that the organism identifies as its own as opposed to all the rest, which is designated as "foreign" or "nonself." This definition does not at all imply that two organisms are necessarily unique: if two strictly identical organisms existed, each of them could certainly identify certain components as "self" and anything else as "nonself."

The arguments above seem to me to illustrate that immunology has failed to determine the ontological question to which it can answer: it claims to respond specifically to the question of identity as *individual description* ("What is this particular living thing?"), most notably as uniqueness ("What makes each living thing unique?"), whereas in reality many biological disciplines address this question, especially those that focus on ontogeny in the broad sense of the living thing's construction throughout its life. I will try to show in the rest of this chapter that the true question that immunology answers is that of identity conceived of as *individuality*; that is, the question of knowing what makes a living thing one discrete and cohesive entity. I will, in other words, show why immunology offers a *criterion of individuation*. Immunology is effectively going to allow us to give a precise definition of the organism and demonstrate that the organism possesses, among all living things, the highest degree of individuality. Individuality, indeed, goes in degrees, and I will try to show that the organism is, among all the living things, that which possesses the highest degree of biological individuality. The major difficulty is that in doing so, I am going to run into the problem laid out in the introduction of this book, that of the confrontation between the criterion of individuation proposed by immunology and the one that currently dominates biology and philosophy of biology—the evolutionary criterion. Given that I am calling into question the dominant conception of what makes a good criterion for biological individuation, it is critical for me to begin with a detailed presentation of the arguments in favor of evolutionary individuation before moving on to make the case for an immunological criterion for individuation. Consequently, I will begin with an explanation of the problem that arises from understanding biological identity as individuality before moving on to develop the evolutionary response to this question and, finally, demonstrating how immunology can explain biological individuality.

IMMUNITY: THE BASIS FOR BIOLOGICAL IDENTITY UNDERSTOOD AS THE ORGANISM'S INDIVIDUALITY

The Issue of Biological Individuality

The question raised by the concept of biological identity understood as individuality is the following: "What counts as an individual in the living world?" The problem of knowing how to individuate biological beings has taken on a considerable importance in contemporary biology and philosophy of biology (Hull 1978; Buss 1987; Sober 1991; Hull 1992; Maynard-Smith and Szathmary 1995; Michod 1999; Santelices 1999; Wilson 1999; Queller 2000; Sober [1993] 2000; Gould 2002; Wilson 2004; Okasha 2006; Gardner and Grafen 2009; Godfrey-Smith 2009; Queller and Strassmann 2009; Folse and Roughgarden 2010). In order to shed some light on these debates, I will start by defining what I mean by "individual" and "biological individual."

What is an *individual*? Every *particular* is not an *individual*, even though every individual is a particular. A *particular* thing is any entity that we may designate by a demonstrative reference (*this F*). An individual is a particular that, moreover, possesses the following characteristics: we can distinguish it (i.e., it is relatively independent, autonomous), count it, it has relatively clear-cut boundaries and a temporal continuity (i.e., it can be said to be the "same" even though it changes through time) (Chauvier 2008). Individuality thus rests on several characteristics, and, moreover, may only be understood as having *degrees*: a given entity will often be *more or less* an individual, in the sense that I am using here. Thus, a stone is more an individual than a pile of sand, a door is more an individual than a doorknob, and a plant is more an individual than a cloud. The word "individual" can refer to natural objects (stones, plants, animals, etc.), as well as artifacts (tables, cars, etc.) but, as we have just seen, not all natural objects or artifacts are individuals; at least, they are not individuals to the same degree. Naturally, it is possible to adopt other definitions of the notion of the individual, but this one is sufficiently broad to account for the majority of responses that, at least since Aristotle, have been raised regarding the problem of defining biological individuality.

What, now, is a *biological individual*? A biological individual is an entity of the living world that possesses the properties of individuality I have just described, which is to say that one can distinguish it (it is relatively independent or autonomous), count it, it has relatively precise boundaries and a temporal continuity.[2] As a result, the term "biological individual" is not a synonym for "organism." The first term is much broader than the second. When we ask what the living individuals are in our world, we do not assume that we are necessarily dealing with "organisms." A gene, a cell, an organ—to take just three examples, could be "biological individuals," without being "organisms." Consequently, like Gould (2002) and many other philosophers and biologists involved in this debate over biological individuality, I reject the a priori equivalence—proposed, for example, by Buss (1987: viii)—between the terms "biological individual" and "organism." According to the definition that I am using, an organism is a biological individual, but all biological individuals are not necessarily organisms. I am also not using the convention chosen by Elliott Sober and David Wilson (Sober and Wilson 1999), which consists of using the term "organism" as a generic one, while "individual" stands for the organism as we intuitively know it. For me, the generic term is "individual," which is effectively the most general one from a metaphysical perspective.

In fact, we can observe that the (more general) question "What is a biological individual?" has, in contemporary biology and philosophy of biology, slowly replaced the question "What is an organism?" previously considered as central (e.g., Huxley 1852; Haeckel 1866; Loeb 1916; Goldstein [1934] 1939; Medawar 1957; Wolvekamp 1966; Lewontin 1983). In this chapter, I attempt to give a new account of biological individuation, based on immunology, but with a specific focus on organisms.

There are three ways of individuating biological identities:

1. *Phenomenal individuation.* The basis for this concept is the idea that one can easily "see" biological individuals. In the same way that a table is considered a good example of an individual in the domain of artifacts, a horse or a cow are considered good examples of biological individuals. The proponents of this concept adopt in fact a commonsense vision of biological individuals.

2. *Physiological individuation.* In this concept, the living world comprises a subclass of biological individuals, that is, organisms, that one can describe as functionally integrated units that change continuously and that are made up of causally interconnected elements. The assumption is that biologists can naturally study other entities, either on a lower level (genes, proteins, tissues, etc.) or on a higher level (groups, species, etc.), but that the fundamental biological individual is the organism, which is conceived of as the only truly unified and autonomous entity in the living world. Kant offers a particularly lucid illustration of this vision (Kant [1790] 2000: §65), which is dominant among physiologists.

3. *Evolutionary individuation.* According to this concept, it is the theory of evolution by natural selection that tells us what a biological individual is. I will return in detail to this idea, but for the moment I will simply say that a biological individual is any entity

upon which natural selection acts. A biological individual may thus be a portion of a genome, a cell, an organism, a species, etc.

In the vast amount of literature dedicated to the question of biological individuation, convincing arguments have been advanced in favor of this third type of individuation. David Hull was one of the best proponents of this concept. In what follows, I shall study Hull's position in detail in order to try to demonstrate its relevance, as well as what I see as its limits. Hull claims that physiology is unfortunately much too vague to allow a satisfactory individuation of living things, and that, therefore, only evolution gives us a satisfactory criterion for individuating living things. Hull played a crucial role since the late 1970s, then, in the way in which philosophers of biology and, to a lesser degree, biologists, have conceived of biological individuality by favoring an evolutionary approach. A criticism of Hull's thesis will show that a truly precise physiological individuation is possible, if it is based on immunology. This is what I shall do after having laid out Hull's arguments. Of course, once the foundations of this physiological individuation are established, it will become necessary to articulate it with evolutionary individuation.

Individuation Based on the Theory of Evolution by Natural Selection

There is a commonsense answer to the problem of biological individuality. To the question "What are the individual entities of the living world?" commonsense responds that organisms are. It is *phenomenal individuation*. The living world seems to be made up of horses, cows, human beings, mosquitoes, and trees—which are described as biological individuals and even, better still, as paradigmatic individuals. According to this concept, we certainly do not have a precise definition of what an organism is, but we do not really need one: we may generally trust our visual perceptions, which tell us what counts as a biological individual.

Yet phenomenal individuation is not at all satisfactory. It would perhaps work for certain organisms we consider "common" (especially domesticated animals), but not for organisms like siphonophores, or slime molds, or many fungi (for more examples, see Huxley 1912; Hull 1988; J. Wilson 1999; R. Wilson 2004; Folse and Roughgarden 2010). Moreover, for a great many plants (strawberry plants, aspens, etc.), commonsense cannot determine where the "biological individual" is located. More often than not, common sense limits itself to examples of familiar organisms around it, forgetting that these make up but a tiny fraction of life. In addition, the problem of biological individuation can be raised even for familiar multicellular organisms. A single cell, for instance, fulfills quite satisfactorily the criteria of individuality described above. Shouldn't we then consider the cell to be a biological individual and the multicellular organism as a "community" comprising cellular biological individuals?

Contrary to common sense, then, it is rather difficult to say with precision which exact individuals constitute the living world. It is generally assumed that one can give an assured response to the question of biological individuation (saying that the living world is composed of organisms); yet, in reality, as soon as we begin to think of specific examples we realize that common sense cannot give a clear, certain response to this question. As Hull (1992) puts it, we cannot trust common sense, because "common sense is strongly biased by our relative size, duration, and perceptual abilities" (Hull 1992; see also Lewontin 2000: 76–77).

If biological individuation cannot be based on perception, on what, then, can it be based? According to Hull (1992), it is our scientific theories in biology, as well as in physics, that tell us which entities make up our world (atoms, fields, genes, etc.). In other words, scientific theories offer an ontology, an image resembling the actual world—or more precisely an image of the individual constituents of which our world is made if we trust what current scientists believe. This does not imply, of course, that the "real world" is as scientists describe it. The idea is simply that theories indicate which individual entities, according to current scientists, comprise our world. Moreover, this scientific ontology is supposedly superior to that of common sense because it allows a better explanation and a better prediction of phenomena. Individuation of entities is thus said to be "theory-dependent."

Now, Hull continues, the only true theory in biology is the theory of evolution by natural selection. The theory of evolution understood as common descent with modification under the effect of natural selection effectively constitutes a theory in the strongest sense of the word, since it unifies various types of independent facts under a single explanatory framework, can be tested empirically, and has largely been corroborated by data from various fields. According to Hull, it is clear that some non-evolutionary fields of biology, such as physiology and morphology, could be very useful for determining what a biological individual is, if only these fields were able to offer a theory. Unfortunately, he says, there is nothing like a physiological or morphological theory, and so only the theory of evolution by natural selection exists to individuate biological entities:

> The trouble with Haeckel's solution to the problem of biological individuals is that morphology and physiology do not provide sufficiently well articulated theoretical contexts. Biologists have been engaged in the study of anatomy and physiology for centuries, but no "theories" of morphology and physiology have materialized in the same sense that evolutionary theory

is a "theory." In order to see the dependence of individuality on theories, one must investigate more highly articulated areas such as evolutionary biology.

(Hull 1992: 184)

If individuation is always dependent on a theory, and if the theory of evolution by natural selection is indeed the only true biological theory, then the best way to individuate biological entities is to determine what an *evolutionary* individual is. In the immense body of literature dedicated to this question in biology and in philosophy of biology, we can observe that, in accordance with Hull's concept, most of the time the search for the definition of the biological individual comes down to determining the definition of the evolutionary individual. Let me examine this concept more closely before moving on to explain where mine differs.

What, then, is an evolutionary individual? The structure of the theory of evolution by natural selection provides the answer to this question. A biological individual is an evolutionary individual, that is to say, any entity upon which natural selection acts (i.e., it is targeted as a whole by natural selection). A biological individual is one "unit of selection" (Lewontin 1970), but in the specific sense of an "interactor" and not a "replicator." What exactly do these terms mean? In order to clarify the debate on units of selection, Hull (1980, 1981, 1988) proposed a distinction at the core of the evolutionary process between the replicator, which is "an entity that passes on its structure largely intact in successive replications" (typically the gene) and the interactor, which is "an entity that interacts as a cohesive whole with its environment in such a way that this interaction causes replication to be differential" (typically the organism). (These definitions are those of Hull [1988]). Whenever the question of evolutionary individuation arises, what is at issue is the interactor (which living entity is natural selection acting upon?), and not the replicator (what entity maintains its structure relatively intact over the evolutionary long run?) (see, e.g., Lloyd 2007; for a critical evaluation of the role played by the "replicator" concept, see Godfrey-Smith 2009).

For now, what is important is to ask what in this world counts as an evolutionary individual and thus as a biological individual. One rather radical possibility is to choose one level of individuality and to hold that this is the only "real" biological level. For example, the thesis of "genic selectionism" holds that the gene is the correct level of biological understanding, on the pretext that a gene persists over time on the evolutionary scale, contrary to an organism (Dawkins 1976, following Williams 1966). In one of his most famous quotes, Dawkins (1976: 34) says, "Genetically speaking, individuals and groups are like clouds in the sky or dust-storms in the desert. They are temporary aggregations or federations. They are not stable through evolutionary time." This concept has been criticized in numerous ways, notably because it does not sufficiently make the distinction possible, at least in some of its formulations, between the interactor and the replicator (see for instance Gould 2002). Others suggest that genic selectionism rests on a confusion between *type* and *token* (see in particular Sober and Lewontin 1982). Be that as it may, genic selectionism may lead to the idea that the living world is, from a scientific point of view, made up of genes and not organisms (Dawkins 1982). Sterelny and Griffiths (1999: 70) even offer a long discussion of the idea that "there is no such thing as an organism."

Yet the most frequently expressed attitude when biological individuals are understood as evolutionary individuals is to support a *hierarchical* conception of evolution (Lewontin 1970; Buss 1987; Maynard-Smith and Szathmary 1995; Gould 1998; Gould and Lloyd 1999; Michod 1999; Queller 2000; Gould 2002; Okasha 2006; Godfrey-Smith 2009; Gardner and Grafen 2009; Queller and Strassmann 2009; Folse and Roughgarden 2010).[3] According to this view, the organism is only one biological individual among others. One cell or one cell lineage may be perfectly legitimate biological individuals because they can be legitimate units of selection (interactors). A good example is adaptive immune cells such as B cells, these cells being, as explained in chapter 1, selected when they encounter an antigen (Burnet 1959; Lewontin 1970; Buss 1987; Michod 1999). Evolutionary individuation thus constitutes a sort of warning to biologists: contrary to what common sense suggests, the organism is not the only biological individual in the world.

Nevertheless, the hierarchical vision of biological individuality goes still further. It leads to a revision of the ontology of the living world. We thought the living world was made up of organisms as we perceive them, but individuation by natural selection shows us that this is simply not true. Daniel Janzen illustrates this attitude particularly well. In a famous article, Janzen (1977) argues that, even though phenomenal and physiological individuations apparently tell us that a dandelion is that thing found in our gardens made of a stem and a flower, evolutionary individuation demonstrates that in reality the true biological individual is the whole clonal dandelion, spread out and with great longevity, since it is to the clone that a "reproductive fitness" can be attributed. This claim rests on the idea that genetically identical organisms like dandelions that reproduce by apomixis cannot be considered as in competition with one another for the transmission of their genes to the next generation. It leads then to the idea that "there may be as few as four individual dandelions competing with each other for the territory of the whole of North America" (Dawkins 1982: 254). The same would go for the aphid evolutionary individual: it is the set of insects that come from one and the same egg and "grow" by parthenogenesis. Given that the insects share the same genome, one cannot say that they are in competition with each other, and one must even say that they constitute "parts" of the same individual (they are like "organs" of one and the same organism).

So evolutionary individuation often finds itself in conflict with the commonsense vision of biological individuation, and leads to a revision of biological ontology. This is likely a good argument in its favor, since it is precisely one of the main functions science performs: it changes the way we see the world with the argument that scientific ontology permits a better explication and prediction of phenomena than does commonsense ontology.

Individuation Based on Physiology

Hull (1992) claims that physiology could be quite useful for defining biological individuality, but considers that it fails to do so. Here, by "physiology," one must understand all *non-evolutionary* biology; that is, the group of biological fields that mainly attempt to answer "how" questions rather than "why" questions. In other words, what I call "physiology" here is what Ernst Mayr (1961) calls "functional biology" (as opposed to evolutionary biology which, for its part, asks "why" questions), but which could more correctly be called "mechanistic biology" to avoid the ambiguity of the term "function" in biology ("functions" in the etiological sense of the word actually belong to "evolutionary biology" in Mayr's sense). Understood this way, the field of physiology includes anatomy, morphology, most of molecular biology (molecular genetics, in particular) and of cellular biology, and most of developmental biology, etc.—but always in their non-evolutionary aspects (for a similar definition of "physiology" or "mechanistic biology," see for instance Neill and Benos 1993; Boron 2005; Pohl 2005).

Why use physiology to define the organism's biological individuality? In other words, what does physiology have to say about biological individuality? As Hull stresses (1974: 75), physiology deals with the organism's functioning and continuous maintenance over time. By seeking to know *how* the organism functions, the goal of the fields that comprise physiology is to explain the organism's construction and maintenance over time; that is, the organism's individuality over time (Neill and Benos 1993; Pohl 2005). This is the classic problem in philosophy of the organism's *sameness*.[4]

Understood this way, it is clear why physiology could become a fundamental tool for comprehending biological individuality: it seems to be a precise means to grasping biological individuation at *one* level of life, that of the organism. Physiology could thus allow us to evaluate the adequacy of commonsense intuition that views the organism as a specific biological individual, that is, the biological entity that probably possesses the highest degree of individuality. This makes the lack of a physiological theory all the more unfortunate if Hull is correct. If, however, he is wrong, then it shows the urgency of proving why a physiological theory is possible and can help lead to a better definition of individuality.

The criterion generally used for offering a physiological individuation is that of *functional integration*. If we think of organisms as different as a plant, a fly, or a rhinoceros, what they all share in common is that they each constitute a functionally integrated and coherent whole (see for instance Sober 1991; Wilson 1999, 2000). In this case, the "natural boundaries," like skin or membranes, are very important, because they help delineate this functionally integrated "whole." This functional integration criterion leads in most cases to the assertion that the best-individuated biological entity is the organism (Kant [1790] 2000; Bernard [1865] 1927; Goldstein [1934] 1939; Gould and Lewontin 1979; Sober 1991; Wilson 1999; Lewontin 2000). Of course, an organ or a "part" in general may also be "functionally integrated," but the idea is that the functional integration reaches a clearly superior degree in the organism taken as a whole.

The problem comes from the fact that the concept of functional integration is too vague to offer a true *criterion* of individuation.

It is too close to phenomenal individuation: we are content to trust our impression that the organism is a coherent "whole," that may be analyzed in functionally defined "parts," and we deduce from that that the individuals in the living world are organisms. But this is a fragile claim. For instance, in the colonial organism *Botryllus schlosseri*, at which biological level is there "functional integration"? Where is the "whole"; where are the "parts"? What are the "natural boundaries" of this organism (De Tomaso et al. 2005)? In *Botryllus schlosseri*, each zoid has a membrane and is, at least to a certain degree, an integrated whole. Yet one could just as easily consider that the true functional integration occurs at the level of the colony, which has a unique vascularization network and a "tunic" that covers all the zoids. Where, in this case, is the biological individual? The same question arises, for example, for the Portuguese man-of-war, which is a marine jellyfish life form that appears to be *one* organism, even though it is actually a colony of specialized organisms. Moreover (as I have already stated), in a multicellular organism, a cell is spatiotemporally localized and functionally integrated: which criteria lead to the claim that the organism is the "true" biological individual—the best-individuated biological entity—in this case?

One solution would be of course to adopt a hierarchical concept of physiological individuation. That would lead to a multiplication of the number of individuals. The cell itself would be an individual in a larger individual (the organism); in *Botryllus schlosseri*, the zoid would be an individual in a larger individual (the colony), and so forth. But in this case, physiology would fail to defend what seems to be one of its main claims, that the organism is the biological individual par excellence. To defend a multilevel concept of physiological individuality does not imply that each of these levels has the same *degree* of individuality. The real question is whether some physiological criteria of biological individuation tend to grant a higher degree of individuality to a specific level among living things.

Functional integration is certainly a good principle, but it needs to be made more precisely founded on a *criterion* of individuation. And to do this, it seems that Hull is correct when he expresses the need for a true physiological *theory*. We must therefore determine if a criterion of individuation based on a physiological theory is possible. In the following section, I suggest that immunology, which is part of physiology as I have defined it here, offers just such a theory of biological individuality.

Individuation Based on a Physiological Theory: Immunity and Biological Individuality

What exact bearing does immunology have on the question of identity understood as individuality? My argument is that immunology strives to offer a criterion of immunogenicity, which is itself a criterion of individuality. As I have pointed out, the immune system, with its surveillance activity, defines what is accepted or rejected by the organism. A criterion of immunogenicity thus constitutes a *criterion of inclusion*: the distinction between entities that are interconnected and form a whole as constituents of the organism and those that are rejected is carried out by the immune system.[5] In other words, immunology allows for an understanding of the living thing's spatial boundaries, and by extension determines which entities constitute its components.

As a result, the immune system is not the same thing as the organism, but it is a subsystem of the organism whose activity leads to the distinction between what is part of the organism and what is not. This distinction has a crucial temporal dimension: for example, a satisfactory criterion of immunogenicity must explain why an organism that possesses a kidney at time t1 can possess, after a transplantation, a second, perfectly tolerated kidney at time t2. Immunity offers a criterion of diachronic inclusion, that is, a criterion of what makes the organism a unit constituted of different entities over time. It resolves the issue of identity as diachronic biological individuality by saying what the organism's components are over time.

A criterion of immunogenicity is thus a particularly precise criterion of individuation, though only, of course, applicable at the level of the *organism* and not for just any living thing. As we shall see when we compare the immunological to the evolutionary criteria of individuation, the first does not claim to have the same extension as the second: whereas evolutionary individuation makes possible to conceptualize a vast hierarchy of individualities (genes, organelles, chromosomes, cells, organism, populations, etc.), immunological individuation concerns itself only with understanding the organism's individuality. It achieves this, though, with a much more important degree of precision, which in turn will make it possible to enrich the evolutionary individuation.

At this point, however, a critical question arises: for exactly which organisms our immunological theory will hold? Is it truly valid for any organism whatsoever? I claim that the theory developed here holds not only for every multicellular organism (plants, invertebrates, vertebrates...) but also for unicellular organisms. Thus, in what follows, after showing that the definition of the organism I suggest holds for all multicellular organisms, I close this chapter with a section in which I explain how my definition of the organism can also be applied to unicellulars.

The notion that the immune system can provide an explanation of what the constituents of an organism are has been expressed intuitively on many occasions. For instance, Gould and Lloyd write: "Organisms are coherently bounded in space and kept recognizable in form by a physical skin that separates the self from the outside world, a distinction often buttressed by various devices—an immune system as the most prominent example—that can recognize and disarm or eliminate transgressors into the interior space" (Gould and Lloyd 1999: 11906). Nevertheless, if the goal is to shed a new light on the issue of biological individuation, we must go beyond intuition (expressed here by Gould and Lloyd) that immunology could help define the organism's individuality, and we need to establish a precise criterion of individuality, founded on a well-defined immunological theory. In effect, immunological individuation depends on the definition of a satisfactory criterion of immunogenicity, which in this case is a criterion of inclusion. I am therefore hoping to build off of Gould and Lloyd's idea by proposing a precise criterion of immunogenicity. The long scientific discussion that threaded through the preceding chapters now becomes critical for the philosophical question of individuality. I hope I have showed the scientific superiority of the continuity theory over the self theory. I would like now to show that the continuity theory offers the precise, experimentally satisfactory criterion of immunogenicity that now allows us to establish the criterion for the individuation of the organism that we have been seeking.

The Continuity Theory and the "Heterogeneous Organism"

According to the continuity theory, there is an immune response whenever there is a strong discontinuity in molecular patterns (ligands) with which immune receptors interact. Any entity that is in a continuous, average-level interaction with immune receptors, or that induces continuity via the mechanism I have described, is tolerated. Continuity theory thus accounts for the tolerance of exogenous entities. I will now explain how the criterion of immunogenicity offered by the continuity theory allows us to come up with a precise physiological definition of the organism.

Before that, however, let us recall that the domain of the continuity theory is vast, since it involves at the very least all multicellular organisms, as well as, in my view, unicellular ones. I ultimately hope to use this theory to establish a general physiological theory of the biological individuation of organisms.

DEFINITION OF THE ORGANISM

I will begin with the usual physiological definition of the organism already discussed: the organism is a functionally integrated whole that undergoes continuous changes and is made up of interconnected elements characterized by causal dependence. Peter's components may interact with those of Paul, but not with the same intensity, regularity, and scale as Peter's components interacting among themselves alone. Although certainly correct, this definition is still too general. Biochemistry can make it more precise. In effect, despite the fact that one may observe the functional integration of the organism on many levels, the actual level where current knowledge tends to recognize it is that of proteins: the different parts of an organism (organs, tissues, cells, and even cell components) are interconnected by strong biochemical interactions that mainly involve protein-protein interactions (Vidal 2001; Venkatesan et al. 2009). In multicellular organisms, a cell than does not receive signals from its local environment and that does not send signals dies quickly. This is true of immune cells, neurons, cells involved in development, etc. Understanding protein interactions is a highly active field in contemporary biology (e.g., Venkatesan et al. 2009). It is the best level to look at in order to understand functional integration within the organism, since the strength, regularity, and extension of "interior" biochemical interactions are quite different from those produced between two different organisms.

Yet even at the biochemical level, functional integration is *local*. In other words, some subsystems in an organism can be quasi-independent (Lewontin 2000: 94). Some very strong biochemical interactions in the liver, for example, may have no influence at all on the brain. At this point, immunology's contribution is decisive: immune interactions involve the entire organism because they are *systemic*. The lymphatic system (or its equivalent) is effectively an extended system that collects extracellular fluid (lymph) in all of the organism's tissues. As a result, all of the organism's tissues and cells are under the influence and control of the immune system.[6]

Immune interactions are therefore a subgroup of biochemical interactions, but first they are *systemic* (not local); second, they offer a *criterion of inclusion* because they are responsible for the acceptance or rejection of components within the organism. Here we reach the heart of the argument. When we associate the biochemical perspective presented above with the immunological perspective that is both specific and systemic, we arrive at the following definition of a multicellular organism:

> *An organism is a functionally integrated whole composed of heterogeneous components that are locally interconnected by strong biochemical interactions and controlled by constant systemic immune interactions of a constant average intensity.*

It is now clear that in my conception, immune interactions are of critical importance and constitute the foundation of this physiological concept of the organism's individuation. First, while biochemical interactions are in most cases local, immune interactions are systemic. Second, whereas it is not always easy to define the intensity of biochemical reactions (due to their diversity), immune interactions are receptor-ligand interactions whose intensity is clearly defined in terms of specificity, affinity, and avidity (see, e.g., Rudolph et al. 2006). Immune cells interact with an average intensity, but not a very strong one, with molecular patterns expressed by the organism's components: if the interactions are very weak, then the immune cell dies; if too strong, it signals that an immune response has been triggered, leading to a possible rejection of the target; it is only when the interactions maintain a regular level of intensity that the organism experiences a normal homeostatic state. These interactions must also be repeated continuously (that is to say regularly: here, "continuous" or "constant" immune reactions refers to the fact that they are repeated in a regular manner, and not that they occur without interruption).

The continuity theory allows then for a precise formulation of an immunological individuation of the organism. This would contradict Hull, showing that a theoretical physiological individuation is possible. The continuity theory is a scientific theory in the strongest sense of the term: it has a well-defined domain, a high level of generality, it unifies distinct classes of phenomena, and it is explicative and predictive. Consequently, it does allow a refutation of Hull's claim that only evolutionary biology offers a true theory upon which to base an understanding of the individuation of living things.

Is immunology the only field within physiology that is up to the task of offering theories? Probably not: other fields of physiology, especially those dealing with molecular biology, can certainly offer theories (developmental biology and cognitive sciences are probably good candidates). It is not certain, however, that many other physiological "theories" in the sense I am using here (i.e., a group of comprehensive, clearly articulated propositions, characterizing a well-defined field, and that are explicatory and predictive) do exist. It is even less certain that these "theories" lead to the establishment of a true *criterion of individuation*. Time will determine whether other fields of physiology besides immunology will advance any precise biological individuation based on a theory.

The definition of the organism I have proposed above does require several additional comments. To begin with, this definition does not imply that anything that does not trigger an immune response in the organism belongs to this very organism: two identical twins may each tolerate the other's organs in the case of transplantation, but that does not imply that the two twins are one and the same organism. On the contrary, my criterion demands both *presence* and *inclusion* (the absence of rejection): an entity is part of the organism only if it undergoes strong biochemical reactions with the rest of the organism (interconnection, presence) and constant systemic immune interactions of an ongoing average intensity with the immune receptors (inclusion).

Second, this definition closely links the state of being an organism with the possession of an immune system. Such an idea can, at first glance, appear surprising. Nevertheless, let us ask the following question: What happens when an organism has no immune system, or only has a very weak one? What are the consequences of *immunodeficiency* for the definition of the organism stated above? It is impossible for an organism to have *no* immunity; it would die immediately. Immunodeficiency is a question of degree. Moreover, it is very often local, with only certain parts of the immune system being affected. If, though, the immune system is massively overcome, if it no longer can perform its surveillance function, then one can say that the organism is on the verge of no longer existing. It is therefore legitimate to link organismicity with immunity.

THE ORGANISM'S HETEROGENEITY

One of the major implications of the definition I am proposing is that it demonstrates that the organism must be considered as "heterogeneous" (following Lewontin 2000). Indeed, my definition stipulates that an organism's components are *heterogeneous*. By this I mean both "different" and "coming from that which is initially outside the organism."[7] Our discussion of immune tolerance phenomena has illustrated the importance of this heterogeneity: a multicellular organism is made up of numerous parts, many of which have not originated in that organism. Put another way, an organism is made of countless foreign entities; it is never constructed in a purely endogenous way. This heterogeneity can be illustrated by the functional role of certain symbiotic bacteria in their host. Each human being is made up of huge quantities of symbiotic bacteria. The majority of these symbiotic bacteria live in the intestine. Some are obligatory symbionts, which means that they cannot survive outside the host, and that without them the host cannot survive either. Many play indispensable physiological (functional) roles, particularly in digestion and immunity. These bacteria carry out permanent and constitutive biochemical interactions with other components of the host. In particular, there is no fundamental difference between interactions between host immune receptors and these symbiotic bacteria, and the interactions between host immune receptors and the host's "own" cells (endogenous, born from the same egg cell): in both cases, there are constant interactions of average intensity, subject to regulatory interactions. This brings me to my main thesis here: these resident bacteria (i.e., the symbiotic bacteria that live in the organism) are not simply "present" in the organism, they are *part* of this organism. One could object that the intestine, which hosts the majority of these symbiotic bacteria, is an interface in the organism rather than a true internal "part." The same objection would be raised for other interfaces in the organism, that is, the other places that have large numbers of bacteria: skin, mouth, lungs, vagina, etc. Yet of the ten organic systems of mammals, eight carry out persistent interactions with normal bacteria (McFall-Ngai 2002), and the same idea holds for virtually all living things in general. To try to exclude environmental interfaces from the definition of the organism is just a new way of asserting the existence of an "interior" in the organism that would be free from any external influence; in reality this is a claim that deprives the organism of a large number—really all—of its functional components. As ecologist Patrick Blandin says, the organism is "a local concentration of interfaces" (personal communication). I have similarly demonstrated that commensal and symbiotic bacteria are not limited to mammals, since they are found in arthropods, plants, colonial organisms, etc.

The notion of including commensal bacteria in the definition of an organism is today increasingly accepted by immunologists and biologists (Lederberg 2000; Hooper and Gordon 2001; Xu and Gordon 2003; Bäckhed et al. 2005; Xu et al. 2007; Round and Mazmanian 2009). O'Hara and Shanahan (2006) characterize intestinal flora as "a forgotten organ," while Gilbert (2002: 212) says that these bacteria "are truly parts of the body" (the same idea is developed further in Gilbert and Epel [2009]). As a result, many specialists have come to share what at first seems to be an audacious and surprising perspective. I hope that the continuity theory offers a theoretical foundation for it, by allowing this perspective to reference more precise arguments than those that have been advanced until now.

Thus any organism is a heterogeneous entity made up of different components having different origins that are unified by common interactions with immune receptors. Consequently, a good criterion of immunogenicity tells us that an organism is first of all a unified whole (its unity is founded on biochemical reactions, and above all on immune interactions), and second, that it is heterogeneous.

BIOLOGICAL GENIDENTITY AND IMMUNE INTERACTIONS
The immunological theory of continuity thus allows the organism to be defined by its immune and biochemical reactions. I would now like to show how the continuity theory revitalizes the ontological thesis of "genidentity."

The genidentity thesis tries to answer the question of a living thing's sameness, which, as we have seen, is the following: "What ensures that a living thing is the same at two different moments, despite changes that affect it between these two moments?" The response the genidentity thesis gives to this question is that individuality over time is ensured by the spatiotemporal continuity of the interactions between components of the being involved. The term *genidentity* is proposed by Hans Reichenbach ([1928] 1958: §21; 1938: chapter 4, §28), who elaborated on Kurt Lewin (1922), but also on John Locke ([1693] 1979), who both advanced this thesis to account for a living thing's individuality over time. The genidentity thesis runs up against a counterargument that has been articulated many times over (see in particular Wiggins 2001): it is impossible to discuss interactions between components without saying *to what* these interactions must be attributed. For how should one isolate, individuate, a group of interactions among all the interactions that present themselves in the world, and say that it is these interactions that, in their continuity, constitute a being? This approach seems to assume the existence a "core" for continuous interactions that are believed to individuate the living thing. The genidentity thesis should thus be rejected in favor of a substantialist conception of individuality over time (Wiggins 2001).

If the organism is defined, however, as the product of continuous biochemical interactions under the control of immune interactions, then the establishment of biological genidentity becomes possible without the need for a substantial "core." In effect, the criterion of immunogenicity provides a criterion that isolates some biochemical interactions as constitutive of a whole, the organism: it is the continuity of biochemical interactions under the control of immune interactions that constitutes the organism. In other words, immune interactions isolate some continuous biochemical interactions, which in turn individuate the organism: the notion of a permanent core of the organism's identity becomes unnecessary.

DIFFERENCE WITH OTHER PHYSIOLOGICAL MEANS OF INDIVIDUATING BIOLOGICAL ENTITIES

The immunological individuation that I am proposing is based on a physiological theory, the continuity theory, contrary to the commonsense notion of individuation and traditional physiological individuation (which rests on the idea of functional integration). What qualifies as an individual in my criterion thus differs from what qualifies as one according to commonsense definitions or traditional physiological individuation. Several examples will serve to illustrate this difference.

To begin with, let us return to the oft-discussed case of the colonial organism *Botryllus schlosseri*. In its case, as we have seen, neither a commonsense concept of individuation nor traditional physiological individuation can tell us if each zoid or the entire colony is a biological individual. My immunological criterion of individuation says that in this case the organism (i.e., the physiological individual) is not each zoid, but rather the colony, which is characterized both by strong biochemical interactions and by a unique immune system (comprising cells like morula cells and phagocytic cells, as well as specific molecules) (Cima et al. 2004), based on a unique histocompatibility complex (this complex, named Fu/HC for *fusion/histocompatibility*, is maintained starting with the larval phase up to the colony phase in *Botryllus*) (De Tomaso et al. 2005). How do we know then, using my immunological criterion, how to delineate the organism in the case of *Botryllus*? Before answering this question, note that when investigating a living thing like *Botryllus*, one is not starting from nothing; one is starting with an entity that appears to be living, but which also does not have a clear spatial beginning or ending. The biochemical criterion alone confirms that there is in fact a living entity, since there are intense biochemical interactions in *Botryllus*, but this does not help delineate the organism either, since these interactions take place within each zoid, as well as in the whole colony. My own immunological criterion demands observation of immune responses in this initially poorly delineated entity. There is immune integration in the entire colony (since it does not trigger an immune response of rejection against the zoids that make it up), whereas each colony does not fuse with every another colony. Immunity thus occurs at the level of the colony, and therefore it is the colony that will count as an organism. To understand this, let us look at the way a *Botryllus* fuses or not with another colony (Litman 2006). *Botryllus* brings up an example of *natural grafting*. Two *Botryllus* colonies that share at least one histocompatibility allele do fuse—which, incidentally, shows that this mechanism cannot be explained by the self-nonself theory (Nyholm et al. 2006). Histo-incompatibility detection seems to be made by the *fester* receptor, which therefore plays a key role in *Botryllus'* immunity (Nyholm et al. 2006). If there is a fusion of two colonies that come into contact, then the result is a *chimera* that contains coexisting genomes of two initially distinct colonies. In certain cases, the stem cells of one of the two colonies become parasites of the other, and in other cases there is simply coexistence (Laird et al. 2005). In all cases, once the fusion has taken place, these two colonies constitute in reality *a single* organism, characterized by its biochemical interactions and by the fact that it triggers unique immune responses.

My criterion also helps respond to the problem of biological individuality for the oft-discussed case of "superorganisms." Many maintain that a colony of social insects (termites, ants, bees, etc.) is more individuated than each of its actual members and should thus be considered an "organism" and the insects as "parts" of this organism (or "superorganism") (Wheeler 1911; Wilson and Sober 1989; Strassman and Queller 2007; Grafen and Gardner 2009; Folse and Roughgarden 2010). Often, however, these arguments mix evolutionary aspects (reproduction conditions, the units of selection issue) and physiological aspects (intensity of interactions between colony members). From this perspective, the term "functional integration" causes problems if it is considered as a characteristic of a "unit of selection" (Wilson and Sober 1989; Sober 1991), since an entity can be functionally very well integrated and yet not constitute a unit of selection, and vice versa. According to my immunological criterion, we must examine what happens when social insects encounter a pathogen such as a bacterium or a fungus. In many cases, the immune response occurs at the level of the individual insect. In the *Zootermopsis angusticollis* termite, for example, the response to the *Pseudomonas aeruginosa* bacterium happens at the insect's level (Rosengaus et al. 1999).

It has recently been suggested, however, that some social insects possess immunity at the colony level too (Cremer, Armitage, and Schmid-Hempel 2007; Cremer and Sixt 2009). This interesting perspective may imply that, in those cases, my immunological criterion will say that the "organism" is the colony rather than each insect. The analogy between immunity at the level of the colony and immunity in the usual sense, however, is often not very refined, as Cremer and Sixt (2009) recognize. Most of the time, immunity is undoubtedly realized at the level of the insect, not the colony. In some cases, colony-level processes facilitate immunity, but it is still an insect-level phenomenon. In a few cases, though, being part of a colony seems to make a critical difference in the capacity to mount an immune response. In the termite *Z. angusticollis*, Traniello et al. (2002) recorded a significantly higher survivorship among nymphs that developed immunity as members of a group in comparison to isolated nymphs. Therefore, a colony-level immunity seems to exist in some species.

According to the view defended here, defining the colony as the "organism" while each insect would be defined as a "part" or "organ" of that organism would require two things. First, there must be strong biochemical interactions between the insects of the colony.

This is usually true with social insects, though in several cases the interactions are not truly "systemic," but rather organized in subsystems (insects interact almost exclusively with insects of the same caste). Second, colony-level immune responses must clearly prevail over insect-level immune responses, and reactions of pathogen rejection must occur mainly at the level of the colony. Some instances appear to meet this condition. For example, some bees have guards that control the nest's entrance and attack or exclude infected nestmates. It seems to be a case of colony-level immunity, and it is in those cases that it appears legitimate to talk about a true "social immunity" (Cremer, Armitage, and Schmid-Hempel 2007). Undoubtedly, more experimental work is needed on these issues. But if it turns out that, in some species, social immunity prevails insect immunity, then there is no doubt that what must be labeled the "organism" is the colony as a whole.

This discussion of social insects also makes clear that the immune-based organismicity as I defined it can exist hierarchically, but then it always comes in degrees. For instance, in a multicellular organism, there exist immune processes triggered at the level of one cell (for instance RNA interference), but the great majority of immune responses result from the many interactions between numerous cells and molecules (including, as we have seen, cells and molecules of "foreign" origin). Consequently, according to the definition given here, the "organism" is the multicellular entity, even though each cell has some degree of organismicity. Following the same criterion in the case of social insects, the organism is in many instances the individual insect, but, in the cases where immunity at the level of the colony prevails immunity at the level of the insect, the organism is the colony, even though the insect has some degree of organismicity.

My criterion of individuation allows the organism to be delineated in cases where common sense cannot. In other cases, my criterion of individuation gives the same result as the commonsense view, but it offers a scientific basis to this conclusion: for example, the criterion says that a mouse we see is effectively an individual organism, but, contrary to common sense, it also tells us what counts as a *part* of the mouse. Counterintuitively, it tells us that entities such as gut bacteria, skin bacteria or long-term tolerated parasites are *parts* of the mouse. Therefore, what is offered here is a true theory of biological individuality leading to ontological revisions or confirmations, as does individuation based on the theory of evolution.

Even in the case where my criterion of immunological individuation converges with commonsense individuation, it does not take the same point of departure. My criterion does not isolate a phenomenally defined organism to attempt to explain it a posteriori: even when it echoes commonsense individuation, it does not at all adopt the same reasoning, since it constitutes for its part a criterion of individuation that can confirm or invalidate the commonsense notion of individuation.

In addition, my criterion shows that the concept that the organism is an endogenous entity, no matter how often it is shared, is erroneous. The idea that the organism is the set of constituents originating from one egg cell, that is, a genetically homogeneous entity, continues to be repeated, even by philosophers who have made major contributions to the problem of biological individuality (e.g., Hull 1978). It has even become a sort of self-evident, unquestioned claim. Yet individuation by immunology shows us that the organism is fundamentally *heterogeneous*. If one accepts my demonstration, then the organism cannot be defined as the set of constituents originating from one egg cell.

I now turn to the question of the articulation between the evolutionary criterion of individuation and the physiological criterion of individuation immunology offers.

Articulating Physiological Individuation with Evolutionary Individuation: The Organism as the Biological Entity Expressing the Highest Degree of Individuality

As we have seen, contemporary biology and philosophy of biology responded to the question of biological individuality almost exclusively with the evolutionary criterion of individuation: Any entity that is acted upon by natural selection can be considered a biological individual. I accept the validity of this criterion, which implies that an entire hierarchy of individuals exists in the living world. This criterion must, however, be articulated with the physiological criterion of individuation. Here I show why this articulation is necessary, and what its consequences are.

One of the implications of the evolutionary criterion of individuation is that the organism is only one level of biological individuality among others (genes, genomes, organelles, cells, populations, species, etc.). In other words, the main biological theory, that is, the theory of evolution by natural selection, would show—contrary to common intuition—that the organism is not the only biological individual, nor a more important level of individuality than any other.

On the contrary, I maintain that the organism is the living thing that possesses the highest degree of individuality. I make this claim with the support of the three following arguments:

1. The heterogeneous organism is defined as a distinct entity with relatively precise boundaries, using the criterion of immunogenicity offered here (the criterion of continuity), which is not the case (at least not in the same degree) with the other levels of individuality in the evolutionary hierarchy.

2. The heterogeneous organism as I have defined it here is in many cases an evolutionary individual. This observation

illustrates why it is necessary to modify the general definition of what counts as an evolutionary individual.

3. The heterogeneous organism exercises control over the variation of lower-level components, particularly cell lineages.

I turn now to these three arguments in detail.

THE HETEROGENEOUS ORGANISM IS A DISTINCT
ENTITY WITH CLEARLY DEFINED BOUNDARIES

As soon as one accepts the idea that evolutionary individuation
must be linked to the physiological individuation that I am proposing, the organism ceases to be a biological individual among
many and becomes what is, undoubtedly, among all evolutionary
individuals, one whose boundaries are better defined. A large part
of Hull's argument claiming that the organism as we perceive it
cannot be a good example of an individual amounts to stating that
its boundaries are not clearly established. This argument is perfectly valid if we are talking about the phenomenal organism. If,
however, one adopts the definition of the organism presented in
this chapter, then the boundaries of the individual organism are
defined with an important degree of precision. The criterion of
individuation presented here does in particular help address problematic cases such as those of colonial organisms like *Botryllus*:
the analysis of the system of immune interactions thus permits
the precise definition of the organism's boundaries. I am certainly
not claiming that my criterion of individuation can account for all
problematic cases of the individuation of the organism, but I do
believe that the best line of reasoning to follow is that of a comparison between different levels of individuality. The organism as
I have defined it has much more clear-cut boundaries than many
other levels of the evolutionary hierarchy. The gene's boundaries,

for example, are difficult to establish (Falk 2000), and we could say the same of a group.

I am not denying that a gene, genome, or a group of organisms may be evolutionary individuals; rather I only wish to highlight that, when we claim that they are individuals, we are stating a very general proposition that does not precisely say what this individual's boundaries are. Lewontin clearly saw this in his article that largely launched the debates on evolutionary individuality (Lewontin 1970), but it has since been forgotten by most. To designate evolutionary individuals is to simply claim that natural selection can act upon a particular biological level. On the contrary, to articulate an evolutionary criterion and a physiological criterion of individuation is to notice that the organism is a relatively well defined and delineated individual.

THE HETEROGENEOUS ORGANISM AS AN EVOLUTIONARY INDIVIDUAL

The immunological-physiological criterion of individuation that I have proposed does have consequences for the hierarchical conception of biological individuality derived from the evolutionary criterion. Clonal organisms, and especially aphids discussed by Janzen (1977) serve to illustrate these consequences. Janzen maintains that during parthenogenesis, the aphid-organism (i.e., the individual insect as we can see it) is not an evolutionary individual. Rather, the evolutionary individual would be the group of all insects originating from a same egg, since they would share the same genome and could not be considered to be in competition with one another. The corresponding idea to Janzen's, more or less inherited from Weismann, is that genetic homogeneity is the key to evolutionary individuality (Buss 1987).

Now, my immunological-physiological criterion suggests something altogether different. Each aphid in the sense of my criterion contains intracellular symbionts whose presence is crucial to the host's survival. These symbionts are transmitted vertically (each aphid transmits its symbionts to its descendents). They are different in each aphid. They can undergo mutations during the aphid's life, modify its fitness and that of its offspring (O'Neill, Hoffmann, and Werren 1997). For example, Dunbar and collaborators have found that a point mutation in the *Buchnera aphidicola* bacteria, whose host is the *Acyrthosiphon pisum* aphid, modifies the host's response to thermal stress, thereby "dramatically affecting host fitness in a manner dependent on thermal environment" (Dunbar et al. 2007). This means that *physiologically* defined parthenogenesis-born aphids are actually in competition with one another: they contain varied endobacteria, which vary from host to host; this variation is inherited and changes the host's fitness. In other words, vertically transmitted bacteria are excellent replicators. Recall that a replicator is "an entity that passes on its structure largely intact in successive replications" (Hull 1988). The example of the aphids, however, demonstrates the need for an extended concept of what counts as a "replicator" (following, in part, Sterelny, Smith, and Dickison 1996). Contrary to Dawkins's, and, to a lesser extent, Hull's claims, genes are not the only replicators in the living world. The question of knowing what the exact entities are that count as replicators remains open, but it is certain that endobacteria are excellent ones (Sterelny 2001).

The definition of the evolutionary individual, especially Janzen's, rests on the question of knowing what the individual transmits to its offspring. If aphids only transmitted their genes to their offspring, then it would be true that the evolutionary individual in the case of the aphid would be the group of all insects originating from the same egg. Whatever the success of each insect would be in terms of survival and reproduction, it would not involve any differential replication of their genes. Everything changes, however, when one realizes that aphids are transmitting far more than their genes, and particularly, that they vertically transmit endobacteria that differ from insect to insect and are replicated differentially.

I believe that my logic of immunological-physiological individuation leads to a major consequence: as the aphid example shows, the argument for genetic homogeneity, central to evolutionary individuation, can lead to false conclusions regarding what counts as an evolutionary individual. In the case of the aphid, in effect, the true evolutionary individual is the aphid as defined by my immunological criterion, and the replicators are not only the aphid's genes, but also the endobacteria it contains.

The argument concerning aphids is likely also valid for most clonal organisms, especially plants, which are hosts to a considerable number of obligatory symbiotic bacteria (Kiers et al. 2003), as well as fungi (Van der Biezen and Jones 1998), even if the mode of transmission (vertical or horizontal) makes a difference. My argument, for instance, probably applies to dandelions. That would constitute a revision of the ontological revision about living things proposed by Janzen: in many clonal organisms, the evolutionary individual would not be the clone, but rather the immunological-physiological organism as I have defined it. As a result, the precise study of physiological mechanisms may bear consequences for what counts as an evolutionary individual.

These analyses lead me to conclude that an evolutionary individual must not be determined based on the sole criterion of genetic homogeneity, since that can lead to an erroneous image of the process of evolution by natural selection. It is necessary, in fact, to closely observe the physiological mechanisms of organisms, especially immunological mechanisms, in order to determine what is or is not an evolutionary individual. Let me be clear here that I am not at all claiming that the heterogeneous organism as I have defined it above is always an evolutionary individual. I am simply suggesting that it is often necessary to start with the heterogeneous organism to determine what the evolutionary individual is, above all in cases where there is vertical transmission of endobacteria.

This conclusion extends the analyses of Buss (1987). Buss used one field of physiology (as I have defined it), namely embryology, to show that the concept of the organism as a genetically homogeneous entity was exact only in a limited number of species (and, even in these very species, only approximately so). He pointed out that many organisms are heterogeneous in the sense that, contrary to Weismann's main idea, their somatic cells can undergo mutations and give birth to germ cells. I hope to extend Buss's thesis here using immunology (another field of physiology) to show that many organisms are heterogeneous in that some of their components do not come from the egg cell and may be transferred to offspring, as well as influence their fitness. In other words, I am suggesting that even the organisms Buss considers "homogeneous" (they would make up part of a small minority of organisms in which Weismannian rules hold true), such as arthropods, are themselves in reality heterogeneous because they comprise entities of different origins that can influence their evolution.

From these analyses, I first deduce that in certain cases the immunological-physiological organism is an evolutionary individual, and that, in all cases, the immunological-physiological organism suggests that to define the evolutionary individual based only on genetic homogeneity is erroneous. Indeed, for those examples considered the most valid in terms of their ability to show that the organism is not a good biological individual (the aphid, for example), I demonstrate that it is in fact an excellent biological individual. More generally, I point out that, even when only considering the evolutionary individual as do Janzen or Dawkins (putting aside the physiological individual), it is still necessary to examine the physiological processes produced in the organism to arrive at a precise definition of what, in each case, counts as an evolutionary individual.

THE HETEROGENEOUS ORGANISM AND THE
CONTROL OF LOWER-LEVEL VARIATIONS

One of the ways to interpret the application of the evolutionary criterion of individuation is to say that it provides a "stacking" concept of levels of individuality. For example, a multicellular organism contains the following levels of individualities: the organism, the cell lineages within this organism, the chromosomes within these cells' nuclei, the genes situated on the chromosomes, etc. Evolutionary individuality is thus expressed as a sort of "Russian doll" of individuals. Choosing the most adequate level of individuality depends therefore on the phenomenon being studied. As we saw, Burnet's clonal selection theory provides a good example.

The idea that biological individuals can exist stacked within each other should not, however, lead to the belief that natural selection acts in the same way upon each level. One of the most important characteristics of the multicellular organism is that, to emerge in the course of evolution, it had to impose a control on variations at lower levels, especially cell lineages. In this section, following (Buss 1987), I am limiting my analysis to multicellular organisms and setting aside unicellular ones. Based on arguments borrowed mainly from embryology, Buss shows that in the course of evolution, the transition from a group of competing cells to a multicellular organism forming a unit necessitated that the latter be capable of limiting the possibility that its constitutive cells replicate themselves at the expense of other cells (Buss 1987: 52–54).

It is crucial to note that the immune system plays a major role in the control of the emergence of variants at the cellular level (Michod 1999). A particularly significant case of dysfunction of this control process is that of cancers (Buss 1987; Michod 1999: 107–32; Frank 2007), where tumor cells proliferate without dying by apoptosis. Normally, the immune system constantly surveys all of the organism's constituents and, in doing so, limits the risk of cell lineages developing uncontrollably. It mobilizes general mechanisms, such as apoptosis induction, but also numerous specific immune cells like natural killer (NK) cells or certain lymphocytes. Moreover, the organism's control over possible variations at a lower level of individuality does not exert only over cell lineages but also over its heterogeneous components like symbiotic and commensal bacteria (Frank 1996), which confirms that our "heterogeneous organism" is the right level for analyzing this phenomenon.

Thus, natural selection can act upon the level of the individual organism only because the organism limits selection upon lineages of lower-level individuals; furthermore, the immune system is one of the principle ways of controlling the emergence of variants on lower levels of individuality. Consequently, the organism is not an evolutionary individual like any other; it is an evolutionary individual that restrains the replicative power of evolutionary individuals contained within it. It is, of course, possible and even probable that natural selection at a higher level than that of the organism—the group or the species, for example—restrains natural selection at the organism level, but it is rare for this control to be exerted with the same regularity and efficacy as that exerted by the organism, thanks to its immune system, on the lower-level entities contained within it.

Based on the three arguments I laid out in this section, the organism as I have defined it is the living thing that possesses the highest level of individuality, since it is physiologically well delineated, it is sometimes itself an evolutionary individual—contrary to what a strictly genetic definition of this notion suggests, and it limits natural selection exerted on some lower levels of individuality to a much higher degree than that which occurs for other levels of living things.

Can the Definition of the Organism Suggested Here Apply to Unicellular Organisms?

Although my goal is to first and foremost propose a satisfactory definition of the multicellular organism, I believe that several arguments exist in favor of extending this definition to unicellular organisms. If we re-examine the main aspects of my demonstration, we can say of a unicellular organism such as a bacterium or archaeon that:

1. It does have a physiology, that is, a functional biology in the sense I have described, characterized in particular by protein-protein interactions. It certainly has a metabolism; prokaryotes carry out metabolic reactions (catabolic and anabolic). It also has many types of enzymes (hydrolase, isomerase, ligase, lyase, oxidoreductase, transferase, etc.) and can replicate its DNA and carry out protein synthesis. Furthermore, the operon model of Jacob and Monod (see, in particular, Jacob and Monod 1961) helped show the complexity of messenger RNA transcription regulations inside a prokaryotic cell (Morange 1998).

2. It does possess an immune system. This "genome immunity" can be based on CRISPR (*clustered regularly interspaced short palindromic repeats*), or other similar interference mechanisms. In addition, I have explained why I think that the continuity theory also applies to unicellular organisms. If I am not mistaken, then, the immunological individuation I have proposed here can be applied to unicellular organisms like bacteria.

3. The question of determining their biological boundaries is worth asking. On the face of it, this question poses fewer difficulties than in multicellular organisms, since the membrane seems to offer a satisfactory delineation. I would like to point out two observations, however. First, the membrane in no way demarcates a clear

and perfectly established boundary: unicellular organisms make constant exchanges with other unicellular organisms, specifically gene exchanges via the pili; fusions sometimes occur; so-called "predatory" bacteria ingest "prey" bacteria. Second, my argument that the organism's identity must not be understood as endogenous is equally valid for single-celled organisms. They, even more than multicellular organisms, permanently integrate material—especially genetic material—that is external. This is especially the case with viral genomes, but also with genes or gene fragments that bacteria exchange between one another (Jain, Rivera, and Lake 1999; Doolittle and Bapteste 2007). These internalized components can become functional: this often occurs with the integration of viral genomes. Recall also that if the reasoning behind the CRISPR system is correct, then it shows that bacteria acquire immunity by integrating fragments of bacteriophage genes, which is without a doubt the best example of functional integration. A bacterium is no more a homogeneous entity nor the product of endogenous development than the multicellular organism is.

Thus, unicellular organisms such as bacteria and archaebacteria are open to the outside, integrating numerous environmental components that come from viruses, other bacteria, etc. It seems reasonable to suppose that the individuality of such a unicellular organism rests on its biochemical interactions, under the control of a mainly genetic immunity. Bacteria effectively internalize in a recurrent manner some foreign material, certain of which acquires a functional role and thus integrates into the bacterium's physiology, but bacteria also reject certain entities with the help of their immune system. They do have, therefore, an immune surveillance mechanism that guarantees the integration or the rejection of exogenous entities. I conclude from these analyses that it is possible to apply my definition of the heterogeneous organism to unicellular organisms.

In this final chapter I have shown that a physiological theory of the individual is possible if it is based on immunology; more exactly, if it is based on a precise criterion of immunogenicity. I have offered a definition of the individual organism based on immune interactions. I have tried to articulate this physiological individuation to the oft-used evolutionary criterion of individuation, and shown what the consequences are of articulating these two modes. Finally, I have attempted to advance a definition of the organism that would be valid for all organisms. If my demonstration is correct, then immunology constitutes a physiological field that is theoretical enough to allow for the suggestion of a precise delineation of the organism; in turn, this definition opens a new inquiry into the recent advances made by biology and philosophy of biology in the study of biological individuation.

CONCLUSION

In this work, I have endeavored to examine some current concepts and theories of immunology. My objective was to determine whether this discipline was capable of giving an answer to the question of what makes a living being's identity. I believe the following theses have been established:

1. The terms "self" and "nonself," though situated at the heart of the discipline, are not defined with sufficient precision to constitute scientific terms. They certainly have a suggestive value in that they bring together a field of experimental biology and moral and psychological preoccupations relative to identity, but I think that such a linkage is ill founded and brings too many risks. I have thus proposed a rejection of the terms "self" and "nonself."

2. The self-nonself theory is experimentally inadequate. Normal autoreactivity and autoimmunity, together with immune tolerance, demonstrate that this theory is not satisfactory and that it would be better to replace it with another.

3. It is necessary to offer a criterion of immunogenicity, that is, a response to the question "in which conditions is an

immune response triggered?" The formulation of any con-current theory to that of the self-nonself must address this requirement.

4. The theory I am proposing, "the continuity theory," is general in its scope (because it concerns a large number of organisms), it is adequate to the available experimental data, and it allows one to establish a certain number of pre-dictions. I submit it to the judgment of immunologists and philosophers.

5. Although immunology cannot, I believe, continue to con-struct itself around the notions of "self" and "nonself," it is not any less capable of responding to the question of what makes the identity of a living being. However, two impor-tant restrictions must be clarified:

    5.1. Immunology only brings a specific and decisive clari-fication to the question of biological *individuality*, and not to other aspects of biological identity such as the uniqueness of the living being or its individual description.

    5.2. Immunology only allows for the definition of individu-ality at the organism level, and not at the level of any living being whatsoever.

6. Immunology offers a physiological theory of individu-ation. In doing so, it takes up David Hull's challenge, which was to find a physiologically based theory, that is, one based on functional biology, which could be used to individuate living beings. Immunology defines the organ-ism as a biological individual. It effectively shows that an organism is a functionally integrated whole, made up of heterogeneous constituents that are locally intercon-nected by strong biochemical interactions and controlled by constant systemic immune interactions of a same

average intensity. By offering a criterion of integration, immunology offers a criterion of individuation, which allows very concretely the establishment of an organism's boundaries.

7. This definition shows that any organism is heterogeneous, partially comprising exogenous entities. This claim corroborates and hopefully strengthens co-constructionist arguments which have been put forth in the last thirty years (Lewontin 1983, 1994, 2000; Oyama [1985] 2000; Gilbert and Epel 2009), according to which the organism is constructed by its environment, as well as it in turn constructs it.

8. When linked with evolutionary individuation, the immunological-physiological individuation that I have proposed demonstrates that the organism is not just any biological individual, but possesses the highest degree of individuality. I base this thesis on three arguments:

   8.1. The heterogeneous organism is defined in a precise manner as a discrete and cohesive entity under the criterion of immunogenicity I have proposed.

   8.2. The heterogeneous organism is in many cases an evolutionary individual. This observation allows us, moreover, to show that it is in certain cases necessary to modify the general definition of what counts as an evolutionary individual.

   8.3. The heterogeneous organism exerts control over the variations of its lower level constituents, in particular its cell lineages.

In this work, evolutionary biology constituted a sort of background, certainly discreet, but permanent (from Burnet and his selective vision of bacteria and later of immune cells, up to the issue of evolutionary individuation). To finish this book by opening up some future perspectives, I hope to make some remarks on the way in which I am situating my work with regard to evolutionary biology.

Evolutionary biology dominates the philosophy of biology worldwide. This is easily explained by the attraction raised by the theory of evolution by natural selection, inarguably the most powerful theory in biology. Furthermore, it is important to note that philosophers of biology that have worked on evolution have figured out how to pose questions both philosophically and scientifically exciting, and have even contributed responses to these questions. These philosophers of biology, of which David Hull or Elliott Sober are two excellent examples, have shown that philosophy can work very closely with science, and that it is not just a matter of asking questions, but that philosophy can, at least sometimes, bring answers.

I do think, however, that the philosophy of mechanistic biology (or physiology) must be developed, particularly the philosophy of molecular biology. The diverse fields of mechanistic biology (or, to use Ernst Mayr's expression, functional biology), those which mainly ask the question "how?," do raise exciting philosophical problems—I hope to have shown this regarding the living being's uniqueness and individuality. In addition, philosophers of mechanistic biology sometimes ask identical questions, or at least similar ones, to those that philosophers of evolutionary biology ask, so much so that, in all these cases, it is useful to connect the two kinds of responses. Here again, the example of biological identity understood as individuality is telling: immunological individuation allows us to suggest a modification to several ideas inherent in evolutionary individuation, and even, in some cases, a modification of the ontology outlined by evolutionary theory.

I am calling for a direct connection between immunology and evolutionary biology, as well as microbiology and ecology. If, in the years to come, such a connection takes place, there is no doubt that new and riveting philosophical questions will emerge.

NOTES

Chapter 2

1. Burnet uses the term "not-self," but the term that was retained by immunologists was "nonself."
2. Medawar, like most scientists practicing transplantation in the 1940s, is a zoologist. It is only in the mid-1950s that immunologists accept transplantation experiments as a field pertinent to their discipline (Brent 1997: 73). In his 1969 work, that is, twenty years after Medawar's main experiments, Burnet must still convince his reader of the fact that transplantation is a phenomenon that falls within the scope of immunology (Burnet 1969: 23–25).

Chapter 3

1. Regulatory T cells are diverse (Shevach 2006). Those that are currently well characterized include Tr1 cells (*type 1 regulatory T cells*), Th3 cells (which produce TGF-β), CD8$^+$ cells with regulatory activity (notably CD8$^+$CD28$^-$, also called "Ts" for "T suppressor cells"), NKT cells with regulatory activity (NK1.1$^+$ T cells), and certain double negative T cells (CD4$^-$CD8$^-$). It is likely that new subpopulations of regulatory cells will be discovered in years to come. One can expect, however, that their classification will change considerably, as the current nomenclature is rather disorganized.

Chapter 4

1. Immune cells do not react with the ligands that are specific to their receptors without interruption; in fact, immune cells interact with them before circulating in the lymphatic system and then interact with them again, etc. The continuity in question here is thus the regular repetition of the same biochemical interaction rather than the maintaining of a single, uninterrupted interaction.

2. I thank Michel Morange for this suggestion.

3. Actually, as Matzinger (1994) notes, it is Lafferty and Cunningham who first proposed the idea that the second signal was caused by an antigen-presenting cell. In Bretscher and Cohn's formulation, another "helper" T cell delivered the second signal.

4. This question is not to be confused with that of the capacity for a phagocytic cell, itself activated, to activate other immune components, leading whether to subsequent tolerance or to destruction of the target (discussed in Green et al. 2009).

5. As I have explained, some extremely rapid mutations leading to equally rapid antigenic changes do not allow a stable biochemical reaction with immune receptors, and can even constitute a pathogenic strategy to evade the immune system. However, the continuity theory predicts that as soon as the expression of an unusual antigenic pattern has been stable for several hours, it induces an immune response.

Chapter 5

1. "Par simple opposition au soi, le seul dénominateur commun au non-soi c'est son caractère transitoire, épisodique, sa confrontation soudaine à un système immunitaire *mature*" (author's italics).

Chapter 6

1. "L'individu, par l'entremise de son système immunitaire, semble avoir une connaissance complète, encyclopédique, de ce qui le constitue."

2. For the purposes of my demonstration, I am setting aside the question, though important and complex, of the definition of life and the delineation between the living and the nonliving. The criteria that are often used (and that I am using here) are chemical activity and exchange (metabolism), organization, and the ability to reproduce with variation. On this question, see Morange (2008).

3. Lewontin and Buss use the term "individual" only in the sense of "multicellular organism," and thus are not using the same vocabulary as I am here. They do, however, discuss "units of selection" and mean by this exactly what I am calling a "biological individual" here.

4. The issue of sameness is raised in particular, as I pointed out in the introduction, by Aristotle (1984a, 1984b), Locke ([1693] 1979), Leibniz ([1765] 1996), Reichenbach (1958), or more recently Wiggins (2001).

5. Of course, other biological activities than that of the immune system lead to the rejection of certain entities. We can think, for instance, of metabolic activities: nutrition (rejection of fecal matter) or even respiration (rejection of CO_2). However, the organism assimilates something and expels the remains of its own assimilative activity with these metabolic activities. In contrast, the immune system accepts or rejects living entities themselves (organs, tissues, bacteria, parasites, etc.) as parts of its identity.

6. In Drosophila, this system is the "hemolymphatic" system. Plants do not have an immune circulatory system. Yet every cell in the plant is capable of synthesizing resistance proteins to pathogenic effectors. In addition, reactions such as programmed cell death show that a form of systemic immune response does exist in plants (DeYoung and Innes 2006).

7. This meaning appears in biology and in medicine: see "heterogeneous" in *The Oxford Dictionary of English* (revised edition), ed. Catherine Soanes and Angus Stevenson (Oxford: Oxford University Press, 2005).

REFERENCES

Adkins, B., Leclerc, C., and Marshall-Clarke, S. 2004. Neonatal adaptive immunity comes of age. *Nature Reviews in Immunology* 4: 553–64.

Alder, M. N., et al. 2005. Diversity and function of adaptive immune receptors in a jawless vertebrate. *Science* 310: 1970–73.

Alizon, S., and van Baalen, M. 2008a. Multiple infections, immune dynamics, and the evolution of virulence. *American Naturalist* 172(4): E150–68.

———. 2008b. Acute or chronic? Within-host models with immune dynamics, infection outcome, and parasite evolution. *American Naturalist* 172(6): E244–56.

Aluvihare, V. R., and Betz, A. G. 2006. The role of regulatory T cells in alloantigen tolerance. *Immunological Reviews* 212: 330–43.

Aluvihare, V. R., Kallikourdis, M., and Betz, A. G. 2004. Regulatory T cells mediate maternal tolerance to the fetus. *Nature Immunology* 5(3): 266–71.

Anderson, C. C. 2009. Placing regulatory T cells into global theories of immunity: An analysis of Cohn's challenge to integrity (Dembic). *Scandinavian Journal of Immunology* 69: 306–9.

Anderson, C. C., and Matzinger, P. 2000a. Danger, the view from the bottom of the cliff. *Seminars in Immunology* 12: 231–38.

———. 2000b. Anderson and Matzinger: Round 2. *Seminars in Immunology* 12: 277–91.

———. 2000c. Anderson and Matzinger: Round 3. *Seminars in Immunology* 12: 331–41.

Anderson, M. S., et al. 2002. Projection of an immunological self shadow within the thymus by the Aire protein. *Science* 298: 1395–401.

Andrian (von), U. H., and Mempel, T. R. 2003. Homing and cellular traffic in lymph nodes. *Nature Reviews in Immunology* 3: 867–78.

Apostolou, I., and von Boehmer, H. 2004. In vivo instruction of suppressor commitment in naive T cells. *Journal of Experimental Medicine* 199(10): 1401–8.

Aristotle. 1984a. *Categories*. In *The complete works of Aristotle*, vol. 1, ed. J. Barnes. Princeton: Princeton University Press.

————. 1984b. *Metaphysics*. In *The complete works of Aristotle*, vol. 2, ed. J. Barnes. Princeton: Princeton University Press.

Ashton-Rickardt, P. G., et al. 1994. Evidence for a differential avidity model of T-cell selection in the thymus. *Cell* 76(4): 651–63.

Atlan, H., and Cohen, I. R., eds. 1989. *Theories of immune networks*. Berlin: Springer-Verlag.

————. 1998. Immune information, self-organization and meaning. *International Immunology* 10(6): 711–17.

Bäckhed, F., et al. 2005. Host-bacterial mutualism in the human intestine. *Science* 307: 1915–20.

Barnes, M. J., and Powrie, F. 2009. Regulatory T cells reinforce intestinal homeostasis. *Immunity* 31: 401–11.

Barrangou, R., et al. 2007. CRISPR provides acquired resistance against viruses in prokaryotes. *Science* 315: 1709–12.

Beg, A. A. 2002. Endogenous ligands of Toll-like receptors: Implications for regulating inflammatory and immune responses. *Trends in Immunology* 23: 509–12.

Belkaid, Y., and Oldenhove, G. 2008. Tuning microenvironments: Induction of regulatory T cells by dendritic cells. *Immunity* 29: 362–71.

Belkaid, Y., and Rouse, B. T. 2005. Natural regulatory T cells in infectious disease. *Nature Immunology* 6(4): 353–60.

Belkaid, Y., and Tarbell, K. 2009. Regulatory T cells in the control of host-microorganism interactions. *Annual Review of Immunology* 27: 551–89.

Belkaid, Y., Blank, R. B., and Suffia, I. 2006. Natural regulatory T cells and parasites: A common quest for host homeostasis. *Immunological Reviews* 212: 287–300.

Belkaid, Y., et al. 2002. CD4⁺CD25⁺ regulatory T cells control *Leishmania major* persistence and immunity. *Nature* 420: 502–7.

Bendelac, A., Savage, P. B., and Teyton, L. 2007. The biology of NKT cells. *Annual Review of Immunology* 25: 297–336.

Benson, A., et al. 2009. Gut commensal bacteria direct a protective immune response against *Toxoplasma gondii*. *Cell Host Microbe* 6: 187–96.

Bernard, C. [1865] 1927. *An Introduction to the study of experimental medicine*. New York: Macmillan Company.

Bernstein, E., et al. 2001. Role for a bidentate ribonuclease in the initiation step of RNA interference. *Nature* 409 (6818): 363–66.

Bettelli, E., Oukka, M., and Kuchroo, V. K. 2007. T_H-17 cells in the circle of immunity and autoimmunity. *Nature Immunology* 8(4): 345–50.

Bianchi, D. W., et al. 1996. Male fetal progenitor cells persist in maternal blood for as long as 27 years post partum. *Proceedings of the National Academy of Sciences USA* 93: 705–7.

Billingham, R. E., Brent, L., and Medawar, P. B. 1953. Actively acquired tolerance of foreign cells. *Nature* 172: 603–6.

Bindea, G., et al. 2010. Natural immunity to cancer in humans. *Current Opinion in Immunology* 22: 215–22.

Bingaman, A. W., et al. 2000. Vigorous allograft rejection in the absence of danger. *Journal of Immunology* 164: 3065–71.

Bischoff, V., et al. 2006. Downregulation of the *Drosophila* immune response by peptidoglycan-recognition proteins SC1 and SC2. *PLoS Pathogens* 2(2): e14, 0139–47.

Bluestone, J. A., and Abbas, A. K. 2003. Natural versus adaptive regulatory T cells. *Nature Reviews in Immunology* 3: 253–57.

Boller, T. 1995. Chemoperception of microbial signals in plant cells. *Annual Review in Plant Physiology and Plant Molecular Biology* 46: 189–214.

Boller, T., and He, S. Y. 2009. Innate immunity in plants: An arms race between pattern recognition receptors in plants and effectors in microbial pathogens. *Science* 324: 742–44.

Bonneville, M., O'Brien, R. L., and Born, W. K. 2010. $\gamma\delta$ T cell effector functions: A blend of innate programming and acquired plasticity. *Nature Reviews in Immunology* 10: 467–78.

Boron, W. F. 2005. Physiology...On Our First Anniversary. *Physiology* 20: 212.

Bouneaud, C., Kourilsky, P., and Bousso, P. 2000. Impact of negative selection on the T cell repertoire reactive to a self-peptide: A large fraction of T cell clones escapes clonal deletion. *Immunity* 13: 829–40.

Bourtzis, K., and Miller, T. A. 2008. *Insect symbiosis*. Boca Raton, Fla.: CRC Press.

Breinl, F., and Haurowitz, F. 1930. Chemische Untersuchungen des Präzipitates aus Hämoglobin und anti-Hämoglobin Serum und Bemerkungen über die Natur der Antikörper. *Hoppe-Seyler Zeitungschrift* 192: 45–57.

Brennan, C. A., and Anderson, K. V. 2004. Drosophila: The genetics of innate immune recognition and response. *Annual Review of Immunology* 22: 14.1–14.27.

Brent, L. B. 1997. *A history of transplantation immunology*. San Diego and London: Academic Press.

———. 2001. Tolerance revisited. In *Singular selves: Historical and contemporary debates in immunology*, ed. A.-M. Moulin and A. Cambrosio, 44–52. Paris: Elsevier.

Bretscher, P., and Cohn, M. 1968. Minimal model for the mechanism of antibody induction and paralysis by antigen. *Nature* 220: 444–48.

———. 1970. A theory of self-nonself discrimination. *Science* 169: 1042–49.

Brownlie, J. C., and Johnson, K. N. 2009. Symbiont-mediated protection in insect hosts. *Trends in Microbiology* 17(8): 348–54.

Burnet, F. M. 1929. 'Smooth-Rough' variation in bacteria in its relation to bacteriophage. *Journal of Pathology and Bacteriology* 32: 15–42.

———. 1940. *Biological aspects of infectious disease*. New York: Macmillan.

———. 1941. *The production of antibodies*. Melbourne: Macmillan.

———. 1957. A modification of Jerne's theory of antibody production using the concept of clonal selection. *Australian Journal of Science* 20: 67–69.

———. 1959. *The clonal selection theory of acquired immunity*. Cambridge: Cambridge University Press.

———. 1960. Immunological recognition of self. *Nobel Lectures in Physiology or Medicine* 3: 689–701.

———. 1962. *The integrity of the body: A discussion of modern immunological ideas*. Cambridge, Mass.: Harvard University Press.

———. 1967. The impact of ideas on immunology. *Cold Spring Harbor Symposia on Quantitative Biology* 32: 1–8.

———. 1968. *Changing patterns: An atypical autobiography*. Melbourne: Heinemann.

———. 1969. *Cellular immunology: Self and notself*. Cambridge: Cambridge University Press.

———. 1970. *Immunological surveillance*. Oxford: Pergamon.

———. 1971. 'Self-recognition' in colonial marine forms and flowering plants in relation to the evolution of immunity. *Nature* 232: 230–35.

———. 1976a. *Immunology, aging and cancer: Medical aspects of mutation and selection*. San Francisco: Freeman and Co.

———, ed. 1976b. *Immunology: Readings from Scientific American*. San Francisco: W. H. Freeman.

Burnet, F. M., and Fenner, F. 1949. *The production of antibodies*, 2d ed. Melbourne: Macmillan.

Buss, L. 1987. *The evolution of individuality*. Princeton: Princeton University Press.

Capron A., et al. 2005. Schistosomes: The road from host–parasite interactions to vaccines in clinical trials. *Trends in Parasitology* 21(3): 143–49.

Carosella, E. D., et al. 2008. HLA-G: From biology to clinical benefits. *Trends in Immunology* 29(3): 125–32.

Cash, H. L., et al. 2006. Symbiotic bacteria direct expression of an intestinal bactericidal lectin. *Science* 313: 1126–30.

Caumartin, J., et al. 2007. Trogocytosis-based generation of suppressive NK cells. *EMBO Journal* 26(5): 1423–33.

Chauvier, S. 2008. Particuliers, individus et individuation. In *L'Individu: Perspectives contemporaines*, ed. P. Ludwig and T. Pradeu, 11–35. Paris: Vrin.

Chen, T.-C., et al. 2004. Generation of anergic and regulatory T cells following prolonged exposure to a harmless antigen. *Journal of Immunology* 172: 5900–5907.

Chen, W., et al. 2003. Conversion of peripheral CD4⁺CD25⁻ naive T cells to CD4⁺CD25⁺ regulatory T cells by TGF-β induction of transcription factor Foxp3. *Journal of Experimental Medicine* 198(12): 1875–86.

Chen, Y., et al. 1994. Regulatory T cell clones induced by oral tolerance: Suppression of autoimmune encephalomyelitis. *Science* 265(5176): 1237–40.

Chen, Y., et al. 1995. Peripheral deletion of antigen-reactive T cells in oral tolerance. *Nature* 376: 177–80.

Chibani-Chennoufi, S., et al. 2004. Phage-host interaction: An ecological perspective. *Journal of Bacteriology* 186(12): 3677–86.

Chisholm, S. T., et al. 2006. Host-microbe interactions: Shaping the evolution of the plant immune response. *Cell* 124: 803–14.

Chomksy, N., 1964. *Current issues in linguistic theory*. The Hague: Mouton.

Cima, F., Sabbadin, A., and Ballarin, L. 2004. Cellular aspects of allorecognition in the compound ascidian *Botryllus schlosseri*. *Developmental and Comparative Immunology* 28: 881–89.

Claas, F. H. J. 2004. Chimerism as a tool to induce clinical transplantation tolerance. *Current Opinion in Immunology* 16: 578–83.

———. 2005. Transplantation: Changing dogmas in clinical transplantation immunology. *Current Opinion in Immunology* 17: 533–35.

Clark, W. R. 2008. *In defense of self: How the immune system really works*. New York: Oxford University Press.

Claverie, J.-M. 1990. Soi et non-soi: Un point de vue immunologique. In *Soi et non-soi*, ed. J. Bernard, M. Bessis, and C. Debru, 35–53. Paris: Seuil.

Cobbold, S. P., et al. 2006. Immune privilege induced by regulatory T cells in transplantation tolerance. *Immunological Reviews* 213: 239–55.

Cohen, I. R. 1989. Natural Id-anti-Id networks and the immunological homunculus. In *Theories of immune networks*, ed. H. Atlan and I. R. Cohen, 6–12. Berlin: Springer-Verlag.

———. 1992a. The cognitive principle challenges clonal selection. *Immunology Today* 13: 441–44.

———. 1992b. The cognitive paradigm and the immunological homunculus. *Immunology Today* 13: 490–94.

———. 2000a. *Tending Adam's garden: Evolving the cognitive immune self*. San Diego: Academic Press.

———. 2000b. Discrimination and dialogue in the immune system. *Seminars in Immunology* 12: 215–19.

Cohen, I. R., and Werkele, H. 1972. Autosensitization of lymphocytes against thymus reticulum cells. *Science* 176(41): 1324–25.

Cohn, M. 1998a. The self-nonself discrimination in the context of function. *Theoretical Medicine and Bioethics* 19: 475–84.

———. 1998b. A reply to Tauber. *Theoretical Medicine and Bioethics* 19: 495–504.

———. 2010. The evolutionary context for a self–nonself discrimination. *Cellular and Molecular Life Sciences* 67: 2851–62.

Combes, C. 2001. *Parasitism: The ecology and evolution of intimate interactions.* Chicago: University of Chicago Press.

Coombes, J. L., and Maloy, K. J. 2007. Control of intestinal homeostasis by regulatory T cells and dendritic cells. *Seminars in Immunology* 19: 116–26.

Coombes, J. L., and Powrie, F. 2008. Dendritic cells in intestinal immune regulation. *Nature Reviews in Immunology* 8: 435–46.

Coombes, J. L., et al. 2007. A functionally specialized population of mucosal CD103+ DCs induces Foxp3+ regulatory T cells via a TGF-β- and retinoic acid-dependent mechanism. *Journal of Experimental Medicine* 204(8): 1757–64.

Cooper, E.L., ed. 1974. *Invertebrate immunology: Contemporary topics in immunobiology*, vol. 4. New York: Plenum Press.

Cooper, E. 2010a. Evolution of immune systems from self/not-self to danger to artificial immune systems. *Physics of Life Reviews* 7: 55–78.

———. 2010b. Self/not-self, innate immunity, danger, cancer potential. *Physics of Life Reviews* 7: 85–87.

Coutinho, A., et al. 1984. From an antigen-centered, clonal perspective of immune responses to an organism-centered network perspective of autonomous reactivity of self-referential immune systems. *Immunological Reviews* 79: 151–68.

Cremer, S., Armitage, S. A. O., and Schmid-Hempel, P. 2007. Social immunity. *Current Biology* 17: R693–R702.

Cremer, S., and Sixt, M. 2009. Analogies in the evolution of individual and social immunity. *Philosophical Transactions of the Royal Society B* 364: 129–42.

Crist, E., and Tauber, A. I. 1999. Selfhood, immunity, and the biological imagination: The thought of Frank Macfarlane Burnet. *Biology and Philosophy* 15(4): 509–33.

Curtis, A. S. G., Kerr, J., and Knowlton, N. 1982. Graft rejection in sponges. *Transplantation* 33(2): 127–33.

Dakic, A., et al. 2004. Development of the dendritic cell system during mouse ontogeny. *Journal of Immunology* 172: 1018–27.

Dangl, J. L., and Jones, J. D. G. 2001. Plant pathogens and integrated defence responses to infection. *Nature* 411: 826–33.

Dausset, J. 1981. The major histocompatibility complex in man. *Science* 213 (4515): 1469–74.

———. 1990. La Définition biologique du soi. In *Soi et non-soi*, ed. J. Bernard, M. Bessis, and C. Debru, 19–26. Paris: Seuil.

Davis, D. M. 2007. Intercellular transfer of cell-surface proteins is common and can affect many stages of an immune response. *Nature Reviews in Immunology* 7: 238–43.

Dawkins, R. 1976. *The selfish gene*. Oxford: Oxford University Press.

———. 1982. *The extended phenotype*. Oxford: Oxford University Press.

Dedeine, F., et al. 2001. Removing symbiotic *Wolbachia* bacteria specifically inhibits oogenesis in a parasitic wasp. *Proceedings of the National Academy of Sciences USA* 98(11): 6247–52.

De Tomaso, A. W., et al. 2005. Isolation and characterization of a protochordate histocompatibility locus. *Nature* 438(24): 454–59.

DeYoung, B. J., and Innes, R. W. 2006. Plant NBS-LRR proteins in pathogen sensing and host defense. *Nature Immunology* 7: 1243–49.

Di Giacinto, C., et al. 2005. Probiotics ameliorate recurrent Th1-mediated murine colitis by inducing IL-10 and IL-10-dependent TGF-β-bearing regulatory cells. *Journal of Immunology* 174: 3237–46.

Ding, S.-W. 2010. RNA-based antiviral immunity. *Nature Reviews in Immunology* 10: 632–44.

Doolittle, W. F., and Bapteste, E. 2007. Pattern pluralism and the Tree of Life hypothesis. *Proceedings of the National Academy of Sciences USA* 104(7): 2043–49.

Dubois, B., et al. 2005. Oral tolerance and regulation of mucosal immunity. *Cellular and Molecular Life Sciences* 62: 1322–32.

Dunbar, H. E., et al. 2007. Aphid thermal tolerance is governed by a point mutation in bacterial symbionts. *PLoS Biol* 5(5): 1006–15.

Dunn, G. P., Koebel, C. M., and Schreiber, R. D. 2006. Interferons, immunity and cancer immunoediting. *Nature Reviews Immunology* 6: 836–48.

Dunn, G. P., Old, L. J., and Schreiber, R. D. 2004. The immunobiology of cancer immunosurveillance and immunoediting. *Immunity* 21: 137–48.

Dunn, G. P., et al. 2002. Cancer immunoediting: From immunosurveillance to tumor escape. *Nature Immunology* 3(11): 991–98.

Eberl, G., and Lochner, M. 2009. The development of intestinal lymphoid tissues at the interface of self and microbiota. *Mucosal Immunology* 2(6): 478–85.

Ehrlich, P. [1897] 1957. The assay of the activity of diphtheria-curative serum and its theoretical basis. In *The collected papers of Paul Ehrlich*, vol. 2, ed. F. Himmelweit, 107–25. Oxford: Pergamon.

———. 1900. On immunity with special reference to cell life. *Proceedings of the Royal Society* 66: 424–48.

Ehrlich, P., and Morgenroth, J. 1901. Über Hämolysine: fünfte Mittheilung. *Berlin klin Wochenschr* 38, 251–56. In *The collected papers of Paul Ehrlich*, vol. 2, ed. F. Himmelweit, 246–55. Oxford: Pergamon.

Elliott, M. R., et al. 2009. Nucleotides released by apoptotic cells act as a find-me signal to promote phagocytic clearance. *Nature* 461: 282–6.

Engelhorn, M. E., et al. 2006. Autoimmunity and tumor immunity induced by immune responses to mutations in self. *Nature Medicine* 12: 198–206.

Falk, R. 2000. The gene—a concept in tension. In *The concept of the gene in development and evolution: Historical and epistemological perspectives*, ed. P. Beurton, R. Falk, and H.-J. Rhenberger, 317–48. Cambridge: Cambridge University Press.

Fehr, T., and Sykes, M. 2004. Tolerance induction in clinical transplantation. *Transplant Immunology* 13: 117–30.

Fire, A., et al. 1998. Potent and specific genetic interference by double-stranded RNA in *Caenorhabditis elegans*. *Nature* 391: 806–11.

Flor, H. H. 1971. Current status of the gene-for-gene concept. *Annual Review of Phytopathology* 9: 275–96.

Folse, H. J., and Roughgarden, J. 2010. What is an individual organism? A multilevel selection perspective. *Quarterly Review of Biology* 85(4): 447–72.

Frank, S. A. 1996. Host control of symbiont transmission: The transmission of symbionts into germ and soma. *American Naturalist* 148(6): 1113–24.

———. 2007. *Dynamics of cancer: Incidence, inheritance, and evolution*. Princeton and Oxford: Princeton University Press.

Freitas, A. A., and Rocha, B. 1999. Peripheral T cell survival. *Current Opinion in Immunology* 11: 152–56.

Friedl, P., den Boer, A. T., and Gunzer, M. 2005. Tuning immune responses: Diversity and adaptation of the immunological synapse. *Nature Reviews in Immunology* 5: 532–45.

Fuchs, E. J., Ridge, J. P., and Matzinger, P. 1996. Immunological tolerance: Response to A. M. Silverstein. *Science* 272(5267): 1407–8.

Gajewski, T. F. 2007. The expanding universe of regulatory T cell subsets in cancer. *Immunity* 27(2): 185–87.

Gallimore, A., et al. 1998. Induction and exhaustion of lymphocytic choriomeningitis virus-specific cytotoxic T lymphocytes visualized using soluble tetrameric major histocompatibility complex class I–peptide complexes. *Journal of Experimental Medicine* 187(9): 1383–93.

Gallucci, S., Lolkema, M., and Matzinger, P. 1999. Natural adjuvants: Endogenous activators of dendritic cells. *Nature Medicine* 5(11): 1249–55.

Gardner, A., and Grafen, A. 2009. Capturing the superorganism: A formal theory of group adaptation. *Journal of Evolutionary Biology* 22(4): 659–71.

Gasser, S., and Raulet, D. H. 2006. Activation and self-tolerance of natural killer cells. *Immunological Reviews* 214: 130–42.

Gayon, J. 1996. The individuality of the species: A Darwinian theory? From Buffon to Ghiselin, and back to Darwin. *Biology and Philosophy* 11: 215–44.

Ghiselin, M. T. 1974. A radical solution to the species problem. *Systematic Zoology* 23: 536–44.

————. 1987. Species concepts, individuality, and objectivity. *Biology and Philosophy* 2: 127–43.

Gilbert, S. F. 2002. The genome in its ecological context. *Annals of the New York Academy of Science* 981: 202–18.

————. 2010. *Developmental biology*, 9th ed. Sunderland, Mass.: Sinauer Associates.

Gilbert, S. F., and Epel, D. 2009. *Ecological developmental biology: Integrating epigenetics, medicine and evolution.* Sunderland, Mass.: Sinauer Associates.

Godfrey, W. R., et al. 2005. Cord blood CD4⁺CD25⁺-derived T regulatory cell lines express FoxP3 protein and manifest potent suppressor function. *Blood* 105(2): 750–58.

Godfrey-Smith, P. 2009. *Darwinian populations and natural selection.* Oxford: Oxford University Press.

Goldstein, K. [1934] 1939. *The organism: A holistic approach to biology derived from pathological data in man.* New York: American Book Company.

Gordon, J. I., et al. 2005. Extending our view of self: The human gut microbiome initiative. <http://genome.gov/10002154>.

Gottesman, S. 2004. The small RNA regulators of *Escherichia coli*: Roles and mechanisms. *Annual Reviews of Microbiology* 58: 303–28.

Goubier, A., et al. 2008. Plasmacytoid dendritic cells mediate oral tolerance. *Immunity* 29: 464–75.

Gould, S. J. 1998. Gulliver's further travels: The necessity and difficulty of a hierarchical theory of selection. *Philosophical Transactions: Biological Sciences* 353(1366): 307–14.

————. 2002. *The structure of evolutionary theory.* Cambridge, Mass.: Harvard University Press.

Gould, S. J., and Lloyd, E. 1999. Individuality and adaptation across levels of selection: How shall we name and generalize the unit of Darwinism? *Proceedings of the National Academy of Sciences USA* 96(21): 11904–9.

Graca, L., Cobbold, S. P., and Waldmann, H. 2002. Identification of regulatory T cells in tolerated allografts. *Journal of Experimental Medicine* 195(12): 1641–46.

Graca, L., et al. 2005. Dominant tolerance: Activation thresholds for peripheral generation of regulatory T cells. *Trends in Immunology* 26(3): 130–35.

Green, D. R., et al. 2009. Immunogenic and tolerogenic cell death. *Nature Reviews in Immunology* 9: 353–63.

Greenberg, J. T. 1996. Programmed cell death: A way of life for plants. *Proceedings of the National Academy of Sciences USA* 93: 12094–97.

Griffiths, P. E. 2001. Genetic information: A metaphor in search of a theory. *Philosophy of Science* 68(3): 394–412.

Grmek, M. D. 1996. L'âge héroïque: les vaccins de Pasteur. In *L'aventure de la vaccination*, ed. A. M. Moulin, 143–159. Paris: Fayard.

Grossman, Z., and Paul, W. E. 2000. Self-tolerance: Context-dependent tuning of T cell antigen recognition. *Seminars in Immunology* 12: 197–203.

Guillet, J.-G., et al. 1987. Immunological self-nonself discrimination. *Science* 235: 865–70.

Ha, E.-M., et al. 2009. Coordination of multiple dual oxidase–regulatory pathways in responses to commensal and infectious microbes in drosophila gut. *Nature Immunology* 10(9): 949–58.

Haeckel, E. 1866. *Generelle Morphologie der Organismen*. Berlin: Georg Reimer.

Hamburger, J. [1976] 1978. *Discovering the individual*. New York: Norton & Company.

Hemmrich, G., Miller, D. J., and Bosch, T. C. G. 2007. The evolution of immunity: A low-life perspective. *Trends in Immunology* 28(10): 449–54.

Henrickson, S. E., et al. 2008. T cell sensing of antigen dose governs interactive behavior with dendritic cells and sets a threshold for T cell activation. *Nature Immunology* 9(3): 282–91.

Herrath (von), M. G., and Homann, D. 2008. Organ-specific autoimmunity. In *Fundamental immunology*, 6th ed., ed. W. E. Paul, 1331–74. Philadelphia: Lippincott Williams & Wilkins.

Hildemann, W. H., et al. 1979. Immunocompetence in the lowest metazoan phylum: Transplantation immunity in sponges. *Science* 204: 420.

Holt, P. G., et al. 2008. Regulation of immunological homeostasis in the respiratory tract. *Nature Reviews in Immunology* 8: 142–52.

Hooper, L. V. 2004. Bacterial contributions to mammalian gut development. *Trends in Microbiology* 12(3): 129–34.

———. 2005. Resident bacteria as inductive signals in mammalian gut development. In *The influence of cooperative bacteria on animal host biology*, ed. M. J. McFall-Ngai, B. Henderson, and E. G. Ruby, 249–64. Cambridge: Cambridge University Press.

Hooper, L. V., and Gordon, J. I. 2001. Commensal host-bacterial relationships in the gut. *Science* 292: 1115–18.

Hooper, L. V., et al. 2003. Angiogenins: A new class of microbicidal proteins involved in innate immunity. *Nature Immunology* 4(3): 269–73.

Horvath, P., and Barrangou, R. 2010. CRISPR/Cas, the immune system of bacteria and archaea. *Science* 327: 167–70.

Houghton, A. N. 1994. Cancer antigens: Immune recognition of self and altered self. *Journal of Experimental Medicine* 180: 1–4.

Howes, M. 2000. Self, intentionality, and immunological explanation. *Seminars in Immunology* 12: 249–56.

Huang, F.-P., and MacPherson, G. 2001. Continuing education of the immune system: Dendritic cells, immune regulation and tolerance. *Current Molecular Medicine* 1: 457–68.

Hull, D. L. 1974. *Philosophy of biological science.* Englewood Cliffs, N.J.: Prentice-Hall.

———. 1978. A matter of individuality. *Philosophy of Science* 45(3): 335–60.

———. 1980. Individuality and selection. *Annual Review of Ecology and Systematics* 11: 311–32.

———. 1981. Units of evolution: A metaphysical essay. In *The philosophy of evolution*, ed. U. L. Jensen and R. Harré, 23–44. Brighton: Harvester Press.

———. 1988. *Science as a process: An evolutionary account of the social and conceptual development of science.* Chicago: Chicago University Press.

———. 1989. *The metaphysics of evolution.* New York: State University of New York Press.

———. 1992. Individual. In *Keywords in evolutionary biology*, ed. E. Fox-Keller and E. Lloyd, 181–87. Cambridge, Mass.: Harvard University Press.

Hunt, J. S. 2006. Stranger in a strange land. *Immunological Reviews* 213: 36–47.

Huxley, J. 1912. *The individual in the animal kingdom.* New York: Cambridge University Press.

Huxley, Thomas H. 1852. Upon animal individuality. *Proceedings of the Royal Institution* 1: 184–89. Reprinted in M. Foster and W. R. Lankester, eds., *The scientific memoirs of Thomas Henri Huxley* (London: Macmillan, 1892), 146–51.

Innes, R. W. 2011. Activation of plant nod-like receptors: How indirect can it be? *Cell Host & Microbe* 9: 87–89.

Ishii, K. J., and Akira, S. 2006. Innate immune recognition of, and regulation by, DNA. *Trends in Immunology* 27(11): 525–32.

Jacob, F., and Monod, J. 1961. On the regulation of gene activity. *Cold Spring Harbor Symposia on Quantitative Biology* 26: 193–211.

Jain, R., Rivera, M. C., and Lake, J. A. 1999. Horizontal gene transfer among genomes: The complexity hypothesis. *Proceedings of the National Academy of Sciences USA* 96: 3801–6.

Janeway, C. A. 1989. Approaching the asymptote? Evolution and revolution in immunology. *Cold Spring Harbor Symposia on Quantitative Biology* 54: 1–13.

———. 1992. The immune system evolved to discriminate infectious nonself from noninfectious self. *Immunology Today* 13(1): 11–16.

———. 2001. How the immune system protects the host from infection. *Microbes and Infection* 3: 1167–71.

Janeway, C. A., Goodnow, C. C., and Medzhitov, R. 1996. Immunological tolerance: Danger—pathogen on the premises! *Current Biology* 6(5): 519–22.

Janeway, C. A., and Medzhitov, R. 2002. Innate immune recognition. *Annual Review of Immunology* 20: 197–216.

Janeway, C. A., Travers, P., Walport, M., and Shlomchik, M. J. 2005. *Immunobiology: The immune system in health and disease.* New York: Garland Science Publishing.

Janzen, D. H. 1977. What are dandelions and aphids? *American Naturalist* 111(979): 586–89.

Jeannin, P., Jaillon, S., and Delneste, Y. 2008. Pattern recognition receptors in the immune response against dying cells. *Current Opinion in Immunology* 20: 530–37.

Jerne, N. K. 1955. The natural selection theory of antibody formation. *Proceedings of the National Academy of Sciences USA* 41: 849–57.

———. 1967. Waiting for the end. *Cold Spring Harbor Symposia on Quantitative Biology* 32: 591–603.

———. 1974. Towards a network theory of the immune system. *Annales d'immunologie* 125 C: 373–89.

———. 1976. The immune system: A network of lymphocyte interactions. In *The immune system,* ed. F. Melchers and K. Rajewski, 259–66. Berlin: Springer Verlag.

———. 1984. Idiotypic networks and other preconceived ideas. *Immunological Reviews* 79: 5–24.

———. 1985. The generative grammar of the immune system. *Science* 229(4718): 1057–59.

Jiang, H., and Chess, L. 2009. How the immune system achieves self–nonself discrimination during adaptive immunity. *Advances in Immunology* 102: 95–133.

Joffre, O., et al. 2009. Inflammatory signals in dendritic cell activation and the induction of adaptive immunity. *Immunological Reviews* 227(1): 234–47.

Jonuleit, H., et al. 2000. Induction of interleukin 10-producing, nonproliferating CD4[+] T cells with regulatory properties by repetitive stimulation with allogeneic immature human dendritic cells. *Journal of Experimental Medicine* 192: 1213–22.

Kahan, B. D. 2003. Individuality: The barrier to optimal immunosuppression. *Nature Reviews Immunology* 3: 831–8.

Kant, I. [1790] 2000. *Critique of the power of judgement.* Translated by P. Guyer and E. Matthews. Cambridge: Cambridge University Press.

Kappler, J. W., Roehm, N., and Marrack, P. 1987. T cell tolerance by clonal elimination in the thymus. *Cell* 49: 273–80.

Kärre, K. 1985. Role of target histocompatibility antigens in regulation of natural killer activity: A reevaluation and a hypothesis. In *Mechanisms of NK mediated cytotoxicity,* ed. D. Callewert and R. B. Herberman, 81–91. San Diego, Calif.: Academy Press.

Kelly, D., Conway, S., and Aminov, R. 2005. Commensal gut bacteria: Mechanisms of immune modulation. *Trends in Immunology* 26(6): 326–33.

Kelly, D., et al. 2004. Commensal anaerobic gut bacteria attenuate inflammation by regulating nuclear-cytoplasmic shuttling of PPAR-γ and RelA. *Nature Immunology* 5: 104–12.

Khosrotehrani, K., et al. 2004. Transfer of fetal cells with multilineage potential to maternal tissue. *Journal of the American Medical Association* 292(1): 75–80.

Khush, R. V., Leulier, F., and Lemaitre, B. 2002. Pathogen surveillance—The flies have it. *Science* 296: 273–75.

Kiers, T. E., et al. 2003. Host sanctions and the legume-rhizobium mutualism. *Nature* 425: 78–81.

Klein, J. 1982. *Immunology: The science of self-nonself discrimination.* New York: Wiley. Second edition Boston: Blackwell Scientific Publications, 1990.

Kocks, C., et al. 2005. Eater, a transmembrane protein mediating phagocytosis of bacterial pathogens in *Drosophila. Cell* 123: 335–46.

Koebel, C. M., et al. 2007. Adaptive immunity maintains occult cancer in an equilibrium state. *Nature* 450: 903–8.

Kourilsky, P., and Claverie, J.-M. 1986. The peptidic self model: A hypothesis on the molecular nature of the immunological self. *Annales d'immunologie* 137: 3–21.

Kovats, S., et al. 1990. A class I antigen, HLA-G, expressed in human trophoblasts. *Science* 248: 220–23.

Kretschmer, K., et al. 2005. Inducing and expanding regulatory T cell populations by foreign antigen. *Nature Immunology* 6(12): 1219–27.

Kurtz, J., and Armitage, S. A. O. 2006. Alternative adaptive immunity in invertebrates. *Trends in immunology* 27(11): 493–96.

Kurtz, J., and Franz, K. 2003. Evidence for memory in invertebrate immunity. *Nature* 425: 37–38.

Ladyman, J., and Ross, D. 2007. *Every thing must go: Metaphysics naturalized.* New York: Oxford University Press.

Lafferty, K. J., and Cunningham, A. 1975. A new analysis of allogenic interactions. *Australian Journal of Experimental Biology and Medical Science* 53: 27–42.

Laird, D. J., De Tomaso, A. W., and Weissman, I. L. 2005. Stem cells are units of natural selection in a colonial ascidian. *Cell* 123: 1351–60.

Langman, R. E. 1989. *The immune system: Evolutionary principles guide our understanding of this complex biological defense system.* San Diego, Calif.: Academic Press.

Langman, R. E., and Cohn, M. 2000. Editorial Introduction. *Seminars in Immunology* 12(3): 159–62.

Lanier, L., and Sun, J. 2009. Do the terms innate and adaptive immunity create conceptual barriers? *Nature Reviews Immunology* 9: 302–3.

Lederberg, J. 2000. Infectious history. *Science* 288(5464): 287–93.

Lee, H. K., and Iwasaki, A. 2007. Innate control of adaptive immunity: Dendritic cells and beyond. *Seminars in Immunology* 19: 48–55.

Leibniz, G. W. [1765] 1996. *New essays on human understanding.* Translated and edited by P. Remnant and J. Bennett. Cambridge: Cambridge University Press.

Lemaitre, B., and Hoffmann, J. 2007. The host defense of *Drosophila melanogaster*. *Annual Review of Immunology* 25: 697–743.

Lemaitre, B., Reichhart, J.-M., and Hoffmann, J. 1997. *Drosophila* host defense: Differential induction of antimicrobial peptide genes after infection by various classes of microorganisms. *Proceedings of the National Academy of Sciences USA* 94: 14614–19.

Lemaitre, B., et al. 1996. The dorsoventral regulatory gene cassette Spätzle/Toll/ cactus controls the potent antifungal response in Drosophila adults. *Cell* 86: 973–83.

Lemaoult, J., et al. 2004. HLA-G1-expressing antigen-presenting cells induce immunosuppressive CD4⁺ T cells. *Proceedings of the National Academy of Sciences USA* 101(18): 7064–69.

Lemaoult, J., et al. 2007. Immune regulation by pretenders: Cell-to-cell transfers of HLA-G make effector T cells act as regulatory cells. *Blood* 109(5): 2040–48.

Leulier, F., and Royet, J. 2009. Maintaining immune homeostasis in fly gut. *Nature Immunology* 10(9): 936–38.

Levy, O. 2007. Innate immunity of the newborn: Basic mechanisms and clinical correlates. *Nature Reviews in Immunology* 7: 379–90.

Lewin, K. 1922. *Der Begriff der Genese*. Berlin: Springer.

Lewontin, R. 1970. The units of selection. *Annual Review of Ecology and Systematics* 1: 1–18.

———. 1983. The organism as the subject and object of evolution. *Scientia* 118: 63–82.

———. 1994. *Inside and outside: Gene, environment and organism*. Worcester, Mass.: Clark University Press.

———. 2000. *The triple helix: Gene, organism and environment*. Cambridge, Mass.: Harvard University Press.

Ley, R. E., Peterson, D. A., and Gordon, J. I. 2006. Ecological and evolutionary forces shaping microbial diversity in the human intestine. *Cell* 124: 837–48.

Litman, G. W. 2005. Colonial match and mismatch. *Nature* 438(7067): 437–39.

———. 2006. How Botryllus chooses to fuse. *Immunity* 25(1): 13–15.

Litman, G. W., and Cooper, M. D. 2007. Why study the evolution of immunity? *Nature Immunology* 8(6): 547–48.

Lloyd, E. A. 2007. Units and levels of selection. In *The Cambridge companion to the philosophy of biology*, ed. D. L. Hull and M. Ruse, 44–65. Cambridge: Cambridge University Press.

Locke, J. [1693] 1979. *An essay concerning human understanding* (2nd edition). Edited by P. H. Nidditch. Oxford: Clarendon Press & New York: Oxford University Press.

Loeb, J. 1916. *The organism as a whole from a physiochemical viewpoint.* New York: G. P. Putnam's sons.

Loeb, L. 1930. Transplantation and individuality. *Physiological Review* 10: 547–616.

———. 1937. The biological basis of individuality. *Science* 86(2218): 1–5.

———. 1945. *The biological basis of individuality.* Springfield, Ill.: Thomas.

Löwy, I. 1991. The immunological construction of the self. In *Organism and the origins of self*, ed. A. I. Tauber, 3–75. Boston Studies in the Philosophy of Science, no. 129. Dordrecht: Kluwer.

Lu, L.-F., et al. 2006. Mast cells are essential intermediaries in regulatory T-cell tolerance. *Nature* 442: 997–1002.

McFall-Ngai, M. J. 2002. Unseen forces: The influence of bacteria on animal development. *Developmental Biology* 242: 1–14.

Macpherson, A. J., and Harris, N. L. 2004. Interactions between commensal intestinal bacteria and the immune system. *Nature Reviews in Immunology* 4: 478–85.

Makarova, K. S., et al. 2006. A putative RNA-interference-based immune system in prokaryotes: Computational analysis of the predicted enzymatic machinery, functional analogies with eukaryotic RNAi, and hypothetical mechanisms of action. *Biology Direct* 1(7).

Margulis, L. 1970. *Origin of eukaryotic cells.* New Haven: Yale University Press.

Margulis, L., ed. 1991. *Symbiosis as a source of evolutionary innovation: Speciation and morphogenesis.* Cambridge, Mass.: MIT Press.

Margulis, L., and Sagan, D. 2002. *Acquiring genomes: A theory of the origins of species.* New York: Basic Books.

Marks, B. R., et al. 2009. Thymic self-reactivity selects natural interleukin 17–producing T cells that can regulate peripheral inflammation. *Nature Immunology* 10: 1125–32.

Marshak-Rothstein, A. 2006. Toll-like receptors in systemic autoimmune disease. *Nature Reviews in Immunology* 6: 823–35.

Maturana, H. R., and Varela, F. J. 1980. *Autopoiesis and cognition: The realization of the living.* Boston Studies in the Philosophy of Science, no. 42. Boston: D. Reidel.

Matzinger, P. 1994. Tolerance, danger, and the extended family. *Annual Review of Immunology* 12: 991–1045.

———. 1998. An innate sense of danger. *Seminars in Immunology* 10: 399–415.

———. 2002. The danger model: A renewed sense of self. *Science* 296: 301–5.

———. 2003. The real function of the immune system or tolerance and the four D's (danger, death, destruction and distress), online publication: <http://glamdring.ucsd.edu>.

———. 2007. Friendly and dangerous signals: Is the tissue in control? *Nature Immunology* 8: 11–13.

Matzinger, P., and Kamala, T. 2011. Tissue-based class control: The other side of tolerance. *Nature Reviews in Immunology* 11: 221–30.

Mayer, L., and Shao, L. 2004. Therapeutic potential of oral tolerance. *Nature Reviews in Immunology* 4: 407–19.

Maynard-Smith, J., and Szathmary, E. 1995. *The major transitions in evolution.* Oxford and New York: W. H. Freeman Spektrum.

Mayr, E. 1961. Cause and effect in biology. *Science* 134: 1501–6.

———. 1987. The ontological status of species: Scientific progress and philosophical terminology. *Biology and Philosophy* 2: 145–66.

Mazumdar, P. H. 1995. *Species and specificity: An interpretation of the history of immunology.* Cambridge: Cambridge University Press.

Medawar, P. B. 1944. The behavior and fate of skin autografts and skin homografts in rabbits. *Journal of Anatomy* 78: 176–99.

———. 1953. Some immunological and endocrinological problems raised by evolution of viviparity in vertebrates. *Symposia of the Society for Experimental Biology* 7: 320–28.

———. 1957. *The uniqueness of the individual.* Londres: Methuen.

———. 1960. Immunological tolerance. *Nobel Lectures in Physiology or Medicine* 3: 704–15.

Medzhitov, R. 2001. Toll-like receptors and innate immunity. *Nature Reviews in Immunology* 1: 135–45.

Medzhitov, R., and Janeway, C. A. 1997. Innate immunity: The virtues of a nonclonal system of recognition. *Cell* 91: 295–98.

———. 2002. Decoding the patterns of self and nonself by the innate immune system. *Science* 296: 298–300.

Melamed, D., et al. 1998. Developmental regulation of B lymphocyte immune tolerance compartmentalizes clonal selection from receptor selection. *Cell* 92: 173–82.

Mellor, A. L., and Munn, D. H. 2004. Policing pregnancy: Tregs help keep the peace. *Trends in Immunology* 25: 563–65.

———. 2006. Immune privilege: A recurrent theme in immunoregulation? *Immunological Reviews* 213: 5–11.

Metchnikoff, E. 1884. Researches on the intracellular digestion of invertebrates. *Quarterly Journal of Microscopical Science* 24: 89–111.

———. [1892] 1893. *Lectures on the comparative pathology of inflammation.* Translated by F. A. Starling and E. H. Starling. London: Kegan, Pauls Trench Trubner.

———. [1901] 1905. *Immunity in infective diseases*. Translated by F. G. Binnie. Cambridge: Cambridge University Press.

Michel, T., et al. 2001. Drosophila Toll is activated by Gram-positive bacteria through a circulating peptidoglycan recognition protein. *Nature* 414: 756–9.

Michod, R. E. 1999. *Darwinian dynamics. Evolutionary transitions in fitness and individuality*. Princeton: Princeton University Press.

Miller, J. F. A. P. 1962. Effect of neonatal thymectomy on the immunological responsiveness of the mouse. *Proceedings of the Royal Society of London* 156B: 410–28.

Moalem, G., et al. 1999. Autoimmune T cells protect neurons from secondary degeneration after central nervous system axotomy. *Nature Medicine* 5: 49–55.

Moffet-King, A. 2002. Natural killer cells and pregnancy. *Nature Reviews in Immunology* 2: 656–63.

Morange, M. 1998. *A history of molecular biology*. Cambridge, Mass.: Harvard University Press.

———. 2005. The ambiguous place of structural biology in the historiography of molecular biology. Symposium *History and Epistemology of Molecular Biology and Beyond: Problems and Perspectives*. Max Planck Institute for the History of Science, October 2005.

———. 2008. *Life explained*. Translated by M. Cobb and M. DeBevoise. New Haven: Yale University Press.

Moulin, A. M. 1990. La Métaphore du soi et le tabou de l'auto-immunité. In *Soi et non-soi*, ed. J. Bernard, M. Bessis, and C. Debru, 55–68. Paris: Seuil.

———. 1991. *Le Dernier Langage de la médecine: Histoire de l'immunologie de Pasteur au Sida*. Paris: PUF.

Murphy, J. B. 1913. Transplantability of tissues to the embryo of foreign species. *Journal of Experimental Medicine* 17: 482–92.

Neill, J. D., and Benos, D. J. 1993. Relationship of molecular biology to integrative physiology. *News in Physiological Sciences* 8: 233–35.

Nossal, G. J. V. 1989. Immunologic tolerance: Collaboration between antigen and lymphokines. *Science* 245(4914): 147–53.

———. 1991. Molecular and cellular aspects of immunologic tolerance. *European Journal of Biochemistry* 202: 729–37.

Noverr, M. C., and Huffnagle, G. B. 2004. Does the microbiota regulate immune responses outside the gut? *Trends in Microbiology* 12(12): 562–68.

Nurnberger, T., et al. 2004. Innate immunity in plants and animals: Striking similarities and obvious differences. *Immunological Reviews* 198: 249–66.

Nyholm, S. V., et al. 2006. Fester: A candidate allorecognition receptor from a primitive chordate. *Immunity* 25: 163–73.

O'Hara, A. M., and Shanahan, F. 2006. The gut flora as a forgotten organ. *EMBO Reports* 7(7): 688–93.

O'Neill, S. L., et al. 1997. *Influential passengers: Inherited microorganisms and arthropod reproduction.* New York: Oxford University Press.

Okasha, S. 2006. *Evolution and the levels of selection.* New York: Oxford University Press.

Oldroyd, G. E. D., Harrison, M. J., and Paszkowski, U. 2009. Reprogramming plant cells for endosymbiosis. *Science* 324: 753–54.

Owen, R. D. 1945. Immunogenetic consequences of vascular anastomoses between bovine twins. *Science* 102: 400–405.

Oyama, S. [1985] 2000. *The ontogeny of information.* Durham, N.C.: Duke University Press.

Pacholczyk, R., et al. 2007. Nonself-antigens are the cognate specificities of Foxp3$^+$ regulatory T cells. *Immunity* 27: 493–504.

Palmer, C., et al. 2007. Development of the human infant intestinal microbiota. *PLoS Biology* 5(7): 1556–73.

Pancer, Z., et al. 2004. Somatic diversification of variable lymphocyte receptors in the agnathan sea lamprey. *Nature* 430: 174–80.

Pardoll, D. 2003. Does the immune system see tumors as foreign or self? *Annual Review of Immunology* 21: 807–39.

Parker, A. S., et al. 2010. Optimization algorithms for functional deimmunization of therapeutic proteins. *BMC Bioinformatics* 11:180.Paul, P., et al. 1998. HLA-G expression in melanoma: A way for tumor cells to escape from immunosurveillance. *Proceedings of the National Academy of Sciences USA* 95: 4510–15.

Pauling, L. 1940. A theory of the structure and process of formation of antibodies. *Journal of the American Chemical Society* 62: 2643–57.

Plasterk, R. H. A. 2002. RNA silencing: The genome's immune system. *Science* 296: 1263–65.

Pohl, U. 2005. Physiology without borders. *Physiology* 20: 148.

Pradeu, T. 2009. Immune system: "Big Bang" in question. *Science* 325: 393.

Pradeu, T., and Carosella, E. D. 2004. Analyse critique du modèle immunologique du soi et du non-soi et de ses fondements métaphysiques implicites. *Comptes rendus de l'Académie des Sciences Biologies* 327: 481–92.

———. 2006a. The self model and the conception of biological identity in immunology. *Biology and Philosophy* 21(2): 235–52.

———. 2006b. On the definition of a criterion of immunogenicity. *Proceedings of the National Academy of Sciences USA* 103(47): 17858–61.

Qin, S., et al. 1993. "Infectious" transplantation tolerance. *Science* 259(5097): 974–77.

Queller, D. C. 2000. Relatedness and the fraternal major transitions. *Philosophical Transactions of the Royal Society B* 355: 1647–55.

Queller, D. C., and Strassmann, J. E. 2009. Beyond society: The evolution of organismality. *Philosophical Transactions of the Royal Society B* 364: 3143–55.

Quine, W. V. O. 1951. Two dogmas of empiricism. *Philosophical Review* 60: 20–43.
———. 1969. Epistemology naturalized. In *Ontological Relativity and Other Essays*. New York: Columbia University Press.

Rairdan, G., and Moffett, P. 2007. Brothers in arms? Common and contrasting themes in pathogen perception by plant NB-LRR and animal NACHT-LRR proteins. *Microbes and Infection* 9: 677–86.

Rakoff-Nahoum, S., et al. 2004. Recognition of commensal microflora by Toll-like receptors is required for intestinal homeostasis. *Cell* 118: 229–41.

Ramet, M. 2001. *Drosophila* scavenger receptor CI is a pattern recognition receptor for bacteria. *Immunity* 15: 1027–38.

Raulet, D. H., Vance, R. E., and McMahon, C. W. 2001. Regulation of the natural killer cell receptor repertoire. *Annual Review of Immunology* 19: 291–330.

Reichenbach, H. [1928] 1958. *The philosophy of space and time*. Translated by M. Reichenbach and J. Freund. New York: Dover Publications.
———. 1938. *Experience and prediction: An analysis of the foundations and the structure of knowledge*. Chicago: University of Chicago Press.

Rescigno, M., et al. 2001. Dendritic cells express tight junction proteins and penetrate gut epithelial monolayers to sample bacteria. *Nature Immunology* 2: 361–67.

Ridge, J. P., Fuchs, E. J., and Matzinger, P. 1996. Neonatal tolerance revisited: Turning on newborn T cells with dendritic cells. *Science* 271(5256): 1723–26.

Rosengaus, R. B., et al. 1999. Immunity in a social insect. *Naturwissenschaften* 86: 588–91.

Rouas-Freiss, N., et al. 1997. Direct evidence to support the role of HLA-G in protecting the fetus from maternal uterine natural killer cytolysis. *Proceedings of the National Academy of Sciences USA* 94, 11520–25.

Rouas-Freiss, N., et al. 2007. Tolerogenic functions of human leukocyte antigen G: From pregnancy to organ and cell transplantation. *Transplantation* 84(1 Suppl): S21–S25.

Round, J. L., and Mazmanian, S. K. 2009. The gut microbiota shapes intestinal immune responses during health and disease. *Nature Reviews in Immunology* 9: 313–23.

Rudolph, M. G., Stanfield, R. L., and Wilson, I. A. 2006. How TCRs bind MHCs, peptides, and coreceptors. *Annual Review of Immunology* 24: 419–66.

Ryals, J. A., et al. 1996. Systemic acquired resistance. *Plant Cell* 8: 1809–19.

Saha, N. R., Smith, J., and Amemiya, C. T. 2010. Evolution of adaptive immune recognition in jawless vertebrates. *Seminars in Immunology* 22: 25–33.

Sakaguchi, S. 2005. Naturally arising Foxp3-expressing $CD25^+CD4^+$ regulatory T cells in immunological tolerance to self and non-self. *Nature Immunology* 6: 345–52.

———. 2006. Regulatory T cells: Meden agan. *Immunological Reviews* 212: 5–7.

Saleh, C., et al. 2009. Antiviral immunity in Drosophila requires systemic RNA interference spread. *Nature* 458: 346–51.

Santelices, B. 1999. How many kinds of individuals are there? *Trends in Ecology and Evolution* 14(4): 152–55.

Sapp, J. 1994. *Evolution by association: A history of symbiosis.* New York: Oxford University Press.

Savill, J., et al. 2002. A blast from the past: Clearance of apoptotic cells regulates immune responses. *Nature Reviews in Immunology* 2: 965–75.

Schmidt, O., Theopold, U., and Beckage, N. E. 2008. Insect and vertebrate immunity: Key similarities versus differences. In *Insect Immunology,* ed. E. Beckage. San Diego: Academic Press.

Scofield, V. L., et al. 1982. Protochordate allorecognition is controlled by a MHC-like gene system. *Nature* 295: 499–502.

Shao, F., et al. 2003. Cleavage of Arabidopsis PBS1 by a bacterial type III effector. *Science* 301: 1230–33.

Shevach, E. M. 2006. From vanilla to 28 flavors: Multiple varieties of T regulatory cells. *Immunity* 25: 195–201.

———. 2009. Mechanisms of Foxp3⁺ T regulatory cell-mediated suppression. *Immunity* 30: 636–45.

Silverstein, A. M. 1985. A history of theories of antibody formation. *Cellular Immunology* 91: 263–83.

———. 1989. *A history of immunology.* New York: Academic Press.

———. 1991. The dynamics of conceptual change in twentieth century immunology. *Cellular Immunology* 132: 515–31.

———. 1996. Immunological tolerance. *Science* 272: 1405.

———. 1999. Paul Ehrlich's passion: The origins of his receptor immunology. *Cellular Immunology* 194: 213–21.

———. 2001. Autoimmunity versus horror autotoxicus: The struggle for recognition. *Nature Immunology* 2: 279–81.

———. 2003. Darwinism and immunology: From Metchnikoff to Burnet. *Nature Immunology* 4(1): 3–6.

Silverstein, A. M., and Rose, N. R. 1997. On the mystique of the immunological self. *Immunological Reviews* 159: 197–206.

Simpson, E. 2006. A historical perspective on immunological privilege. *Immunological Reviews* 212: 12–22.

Smith, D. W., and Nagler-Anderson, C. 2005. Preventing intolerance: The induction of nonresponsiveness to dietary and microbial antigens in the intestinal mucosa. *Journal of Immunology* 174: 3851–57.

Smits, H. H., et al. 2005. Different faces of regulatory DCs in homeostasis and immunity. *Trends in Immunology* 26(3): 123–29.

Snell, G. D. 1948. Methods for the study of histocompatibility genes. *Journal of Genetics* 49: 87–108.

Sober, E. 1991. Organisms, individuals and units of selection. In *Organism and the origins of self*, ed. A. I. Tauber, 275–96. Boston Studies in the Philosophy of Science, no. 129. Dordrecht: Kluwer.

———. [1993] 2000. *Philosophy of biology*. Boulder, Colo.: Westview Press.

Sober, E., and Lewontin, R. C. 1982. Artifact, cause and genic selection. *Philosophy of Science* 49(2): 157–80.

Sober, E., and Wilson, D. S. 1999. *Unto others: The evolution and psychology of unselfish behavior*. Cambridge, Mass.: Harvard University Press.

Somerset, D. A., et al. 2004. Normal human pregnancy is associated with an elevation in the immune suppressive CD25$^+$ CD4$^+$ regulatory T-cell subset. *Immunology* 112: 38–43.

Sontheimer, E. J., and Carthew, R. W. 2005. Silence from within: Endogenous siRNAs and mi RNAs. *Cell* 122(1): 9–12.

Spörri, R., and Reis e Sousa, C. 2005. Inflammatory mediators are insufficient for full dendritic cell activation and promote expansion of CD4$^+$ T cell populations lacking helper function. *Nature Immunology* 6(2): 163–70.

Steinman, R. M., and Nussenzweig, M. C. 2002. Avoiding horror autotoxicus: The importance of dendritic cells in peripheral T cell tolerance. *Proceedings of the National Academy of Sciences USA* 99(1): 351–58.

Steinman, R. M., Hawiger, D., and Nussenzweig, M. C. 2003. Tolerogenic dendritic cells. *Annual Review of Immunology* 21: 685–711.

Steinman, R. M., et al. 2000. The induction of tolerance by dendritic cells that have captured apoptotic cells. *Journal of Experimental Medicine* 191(3): 411–16.

Sterelny, K. 2001. Niche construction, developmental systems, and the extended replicator. In *Cycles of contingency: Developmental systems and evolution*, ed. S. Oyama, P. E. Griffiths, and R. D. Gray, 333–49. Cambridge, Mass.: MIT Press.

Sterelny, K., and Griffiths, P. E. 1999. *Sex and death: An introduction to philosophy of biology*. Chicago: University of Chicago Press.

Sterelny, K., Smith, K. C., and Dickison, M. 1996. The extended replicator. *Biology and Philosophy* 11: 377–403.

Stoner, D. S., Rinkevich, B., and Weissman, I. 1999. Heritable germ and somatic cell lineage competitions in chimeric colonial protochordates. *Proceedings of the National Academy of Sciences USA* 96: 9148–53.

Strassman, J. E., and Queller, D. C. 2007. Insect societies as divided organisms: The complexities of purpose and cross-purpose. *Proceedings of the National Academy of Sciences USA* 104: 8619–26.

Strawson, P. F. 1959. *Individuals: An essay in descriptive metaphysics*. London: Methuen & Co.

Strober, W. 2009. The multifaceted influence of the mucosal microflora on mucosal dendritic dell responses. *Immunity* 31: 377–88.

Stuart, L. M., and Ezekowitz, R. A. B. 2005. Phagocytosis: Elegant complexity. *Immunity* 22: 539–50.

Suber, F., Carroll, M. C., and Moore, Jr., F. D. 2007. Innate response to self-antigen significantly exacerbates burn wound depth. *Proceedings of the National Academy of Sciences USA* 104(10): 3973–7.

Suffia, I. J., et al. 2006. Infected site-restricted Foxp3⁺ natural regulatory T cells are specific for microbial antigens. *Journal of Experimental Medicine* 203: 777–88.

Sykes, M. 2001. Mixed chimerism and transplant tolerance. *Immunity* 14: 417–24.

Tafuri, A., et al. 1995. T cell awareness of paternal alloantigens during pregnancy. *Science* 270: 630–33.

Taiz, L., and Zeiger, E. [1991] 2006. *Plant physiology*, 4th ed. Sunderland: Sinauer Associates.

Talmage, D. W. 1957. Allergy and immunology. *Annual Reviews of Medicine* 8: 239–56.

Tauber, A. I. 1994. *The Immune Self: Theory or Metaphor?* Cambridge: Cambridge University Press.

———. 1997. Historical and philosophical perspectives on immune cognition. *Journal of the History of Biology* 30: 419–40.

———. 1999. The elusive immune self: A case of category errors. *Perspectives in Biology and Medicine* 42(4): 459–74.

———. 2000. Moving beyond the immune self? *Seminars in Immunology* 12: 241–48.

———. 2009 [2002]. The biological notion of self and non-self. *Stanford Encyclopedia of Philosophy* (online).

Tauber, A. I., and Chernyak, L. 1991. *Metchnikoff and the origins of immunology*. New York: Oxford University Press.

Taylor, P. R., et al. 2005. Macrophage receptors and immune recognition. *Annual Review of Immunology* 23: 901–44.

Thomas, L. 1959. Discussion. In *Cellular and humoral aspects of the hypersensitive states*, ed. H. S. Lawrence, 529–32. New York: Hoeber-Harper.

Tonegawa, S. 1983. Somatic generation of antibody diversity. *Nature* 302: 5009–15.

Tonegawa, S., et al. 1974. Evidence for somatic generation of antibody diversity. *Proceedings of the National Academy of Sciences USA* 71(10): 4027–31.

Traniello, J. F. A., Rosengaus, R. B., and Savoie, K. 2002. The development of immunity in a social insect: Evidence for the group facilitation of disease resistance. *Proceedings of the National Academy of Sciences USA* 99: 6838–42.

Turnbaugh, P. J., et al. 2007. The Human microbiome project. *Nature* 449: 804–10.

Vaishnava, S., et al. 2008. Paneth cells directly sense gut commensals and maintain homeostasis at the intestinal host-microbial interface. *Proceedings of the National Academy of Sciences USA* 105:20858–63.

van der Biezen, E. A., and Jones, J. D. G. 1998. Plant disease-resistance proteins and the gene-for-gene concept. *Trends in Biochemical Sciences* 23: 454–56.

van der Oost, J., et al. 2009. CRISPR-based adaptive and heritable immunity in prokaryotes. *Trends in Biochemical Sciences* 34(8): 401–7.

Vance, R. E. 2000. A Copernican revolution? Doubts about the danger theory. *Journal of Immunology* 165: 1725–28.

Varela, F. J., et al. 1988. Cognitive networks: Immune, neural, and otherwise. In *Theoretical immunology* (Part Two), ed. A. S. Perelson, 359–75. Redwood City: Addison-Wesley Publishing Co.

Vaz, N. M., and Varela, F. J. 1978. Self and non-sense: An organism-centered approach to immunology. *Medical Hypotheses* 4: 231–67.

Venkatesan, K., et al. 2009. An empirical framework for binary interactome mapping. *Nature Methods* 6(1): 83–90.

Vidal, M. 2001. A biological atlas of functional maps. *Cell* 104: 333–39.

Vivier, E., and Malissen, B. 2005. Innate and adaptive immunity: Specificities and signaling hierarchies revisited. *Nature Immunology* 6(1): 17–21.

von Boehmer, H. 2003. Dynamics of suppressor T cells: In vivo veritas. *Journal of Experimental Medicine* 198(6): 845–49.

Waldmann, H. 2002. Reprogramming the immune system. *Immunological Reviews* 185(1): 227–35.

Waldmann, H., et al. 2004. Regulatory T cells and organ transplantation. *Seminars in Immunology* 16: 119–26.

Waldmann, H., et al. 2006. Regulatory T cells in transplantation. *Seminars in Immunology* 18: 111–19.

Wang, X.-H., et al. 2006. RNA interference directs innate immunity against viruses in adult Drosophila. *Science* 312: 452–54.

Wells, H. G., Huxley, J. S., and Wells, G. P. [1929] 1934. *The science of life.* New York: Literary Guild.

West-Eberhard, M. J. 2003. *Developmental plasticity and evolution.* Oxford: Oxford University Press.

Wheeler, W. M. 1911. The ant-colony as an organism. *Journal of Morphology* 22: 307–25.

Wiggins, D. [1980] 2001. *Sameness and substance renewed.* Oxford: Basil Blackwell.

Williams, G. C. 1966. *Adaptation and natural selection: A critique of some current evolutionary thought.* Princeton: Princeton University Press.

Wilson, D. 1971. *The science of self: A report of the new immunology.* Essex: Longman.

Wilson, D. S, and Sober, E. 1989. Reviving the superorganism. *Journal of Theoretical Biology* 136: 337–56.

Wilson, J. 1999. *Biological individuality: The identity and persistence of living enti-ties.* Cambridge: Cambridge University Press.

———. 2000. Ontological butchery: Organism concepts and biological general-izations. *Philosophy of Science (Proceedings)* 67: S301–S311.

Wilson, R. 2004. *Genes and the agents of life: The individual in the fragile sciences (Biology).* Cambridge: Cambridge University Press.

Wodarz, D. 2006. Ecological and evolutionary principles in immunology. *Ecology Letters* 9: 694–705.

Wolvekamp, H. P. 1966. The concept of the organism as an integrated whole. *Dialectica* 20: 196–214.

Wood, K. J., and Sakaguchi, S. 2003. Regulatory T cells in transplantation toler-ance. *Nature Reviews Immunology* 3: 199–210.

Wraith, D. C. 2006. Avidity and the art of self non-self discrimination. *Immunity* 25: 191–93.

Xu, J. 2007. Evolution of symbiotic bacteria in the distal human intestine. *PLoS Biology* 5(7): 1574–86.

Xu, J., and Gordon, J. I. 2003. Honor thy symbionts. *Proceedings of the National Academy of Sciences USA* 100(18): 10452–59.

Xu, J., et al. 2004. Message from a human gut symbiont: Sensitivity is a prerequi-site for sharing. *Trends in Microbiology* 12(1): 21–28.

Yokota, T., et al. 2006. Tracing the first waves of lymphopoiesis in mice. *Development* 133: 2041–51.

Yong, Z., et al. 2007. Role and mechanisms of CD4$^+$CD25$^+$ regulatory T cells in the induction and maintenance of transplantation tolerance. *Transplant Immunology* 17: 120–29.

Zaidman-Remy, A., et al. 2006. The *Drosophila* amidase PGRP-LB modulates the immune response to bacterial infection. *Immunity* 24: 463–73.

Zinkernagel, R. M. 2004. Credo 2004. *Scandinavian Journal of Immunology* 60: 9–13.

Zinkernagel, R. M., et al. 1978. On the thymus in the differentiation of "H-2 self-recognition" by T cells: Evidence for dual recognition? *Journal of Experimental Medicine* 147: 882–96.

INDEX

adaptive immunity, 18, 24, 27, 31, 34, 36, 76, 110, 152, 165, 195, 211, 215–6

antibody, 19, 41, 55–6, 62, 65–6, 68, 75, 79–80, 192, 200

antigen, 7, 17, 19–21, 41, 63, 65–6, 69, 79, 85, 87, 89, 91, 94, 96, 102, 106, 116, 128, 131–40, 143–7,152, 155, 158–9, 169–70, 181, 192, 203, 212

antigen presenting cell (APC), 20, 33, 36, 89, 91, 93, 106, 113–4, 119, 135, 140, 153, 156, 167, 206–7

apoptosis, 91–2, 95, 137, 164, 261–2

Aristotle, 3, 229

Atlan, H, 129, 190, 197, 200–4

autoimmunity, 36, 62, 69, 82, 85–8, 90, 94–5, 97–9, 107, 135, 153, 159, 162, 167, 169, 172–3, 176–7, 180, 198–202, 267

and autoreactivity, 82, 85–99, 134–6, 148, 162–70, 177, 191–5, 199, 201–4, 267

autopoiesis, 190, 195, 197–200

B cells: see 'lymphocytes'

bacteriophage, 29–31, 65, 264

Belkaid, Y, 124–5, 153, 155, 157–8, 166–8, 170

Bernard, C, 238

biochemical interactions, 20–1, 86, 135, 150, 196, 199, 215, 243–4, 246–52, 264, 268

Botryllusschlosseri, 74–5, 89, 108, 145, 212, 239,250–1, 256

Brent, L B, 53, 63, 111, 136, 148

Burnet, F M, 5, 8, 10, 12, 22–3, 29, 32, 41, 42, 46–50, 55–77, 81–2, 99–102, 108–11, 172, 188, 213, 236

Buss, L, 3, 9, 228–9, 235–6, 257, 260–1

cancer, 82, 89, 94, 122, 125–6, 132, 137, 167–8, 170–2, 181–2, 213, 261

Carosella, E D, 107–8, 113, 131, 221

chimerism, 60, 63, 102–3, 115–6, 159, 174, 251

Clark, W R, 16, 42, 81

clonal selection theory, 49–50, 64–68, 73–5, 84, 191, 261

clonal organisms, 3, 236, 257, 259

Cohen, I R, 81, 129, 196–7, 200–12, 225

Cohn, M, 81, 152, 206

cognition, and immunity, 195–7, 201, 203–4

continuity theory
  presentation of, 131–146, 180–3, 242
  compared with other theories, 185–218
  induction of continuity, 155–62

colonial organisms, 1, 89, 108, 145, 239, 248, 250, 256
Cooper, E, 23, 25, 131
Coutinho, A, 190, 196, 198, 199
CRISPR, 30–1, 179, 263–4
cytokines, 35–7, 83, 87, 89, 105–6, 108, 115, 141, 160–1, 164–7, 211

danger theory, 82, 190, 205–18
Dausset, J, 6–7, 42, 45, 77–8, 224
Dawkins, R, 235–6, 258, 260
DeTomaso, A W, 89, 108, 212, 239, 250
defense, 4, 15–16, 19, 26, 28, 30, 42, 72, 76, 78, 81, 98, 107, 113, 216
dentritic cells: *see 'antigen presenting cells'*
development of the immune system, 43, 45, 61–4,67, 70–1, 100, 103, 121–3, 140–3, 148–51, 158–9, 225–6

Eberl, G, 121
ecology, 55–58, 68, 71, 75, 124, 270
Ehrlich, P, 49–50, 52–53, 56, 62, 65, 79, 90, 99, 209
evolution, and immunity, 22–25, 74–6, 175–80, 255–62, 269–70

Frank, S A, 261–2

genidentity, 248–9
Gilbert S F, 123, 226, 248, 269
Godfrey-Smith, P, 3, 9, 221, 228, 235
Goldstein, K, 230, 238
Gordon, J I, 117–9, 121–2, 161, 248
Gould, S J, 3, 221, 228, 229, 235, 238, 241, 242
Graft: *see 'transplantation'*
Griffiths, P E, 203, 235

heterogeneity: see 'organism'
Hamburger, J, 5, 8
hierarchical conception of evolution, 9–10, 235–6, 241, 255–7, 261
histocompatibility, 6–7, 45, 77–8, 88–9, 101, 108, 113, 140, 144, 250–1
Hoffmann, J, 23, 29, 38, 93, 95, 143, 177
homeostasis, 86, 120, 122, 135, 162, 164

Hooper, L V, 117–9, 121, 248
horror autotoxicus, 50, 52, 56, 61, 62, 90, 99
Hull, D L, 3, 9, 221, 228, 231–4, 237–8, 240, 245, 254, 256, 258, 268, 270
human leukocyte antigen (HLA), 7, 45, 88–90, 101, 224
    HLA-G, 87, 104–8, 112–5, 125–6, 154, 158, 182
Huxley, J, 42, 57, 58, 232
Huxley, T, 230

identity, 1–13, 51, 52, 54, 219–20, 222–31
idiotype and idiotypic network, 192
immune memory, 7, 17–8, 23–4, 27, 31, 59, 76, 133, 156, 196, 203, 225
immune reaction,vs. immune response, 86–7
immune system, description of, 32–41
immunization, 16–19, 158
immunogenicity, criterion of, 39, 69, 71–72, 83, 134, 143–6, 207, 218, 240–1, 265, 267
immunology, definition of, 15–22
immunoprivileged organs, 104–5
inflammation, 34–5, 120
information, 196, 203
innate immunity, 18, 23, 27, 29, 34, 36–7, 92–4, 97, 109–10, 132, 135, 139, 142–3, 152, 195, 199, 211, 216–7
individuality, 1, 2, 69, 222, 227–42
    evolutionary, 9, 231–7, 255–62
    immunological, 8, 54, 240–55, 268
Innes, R, 27, 39, 40, 98, 174, 176, 177
insect immunity, 23, 25, 37–38, 110–8, 123, 177–8, 236, 251–3, 257–9
integrity, 8, 51, 59, 71, 188, 194
invertebrate immunity, 23–24, 28, 89, 98, 110, 123, 175

Janeway, C A, 16–7, 23, 33, 80, 110, 152, 206–7, 210–1, 214, 216
Janzen, D H, 236, 257–60
Jerne, N K, 65–6, 73–4, 82, 86, 190–201, 204

Kant, I, 230, 238

Koch, R, 17–8
Koonin, E, 30–1
Kourilsky, P, 45, 80, 90, 187–8
Kurtz, J, 24

Ladyman, J, 221
Langman, R E, 81
Leibniz, G W, 3
levels of selection, 3, 235, 255–7, 261–2
Lewontin, R C, 9, 230, 232, 234–6, 238,
    243, 246, 257, 269
ligand, 19–22, 26–7, 86–7, 115, 131–3, 135,
    139, 140, 143, 150, 152–4, 163, 166,
    173, 186–7, 242, 244
Litman, G W, 24, 110, 251
Lloyd, E A, 235, 241–2,
Locke, J, 3, 43, 249
Loeb, J, 230
Loeb, L, 8, 54, 224
Löwy, I, 129
lymphocytes
    B lymphocytes, 35, 36, 152, 206
    T lymphocytes, 35, 36, 78–9, 90–2, 97,
        152, 206, 216

McFall-Ngai, M, 117–8, 123, 247
major histocompatibility complex (MHC)
    see 'histocompatibility' and 'human
    leukocyte antigen'
Margulis, L, 117
Maturana, H R, 190, 196–7
Matzinger, P, 82, 190, 193, 203, 205–18
Maynard-Smith, J, 3, 9, 221, 228, 235
Mayr, E, 117, 221, 237, 270
Medawar, P, 8, 45, 54–5, 60–1, 63–4, 69,
    100–1, 104, 112, 136, 148–9, 189,
    225, 230
medicine and immunology, 32, 53, 76, 111
Medzhitov, R, 20, 23, 110, 211
metaphysics, 4, 42, 220–22, 227, 229, 233,
    236–7, 248, 254, 259, 270
Metchnikoff, E, 18, 49–51, 78, 95, 164, 209
Michod, R E, 3, 9, 221, 228, 235–6, 261
microbiology, 18, 100, 124, 270
molecular biology, 20–1, 31, 45, 82–3,
    86, 135–41, 144–7, 150, 163–4,

173, 180–2, 196, 199–200, 210, 215,
    221, 237, 242–52, 264, 268, 270
Morange, M, 215, 263
Moulin, AM, 17–8, 52, 62, 100, 129, 191
Murphy, J B, 53, 54, 60

natural killer (NK) cells, 34–5, 80, 87, 93,
    164, 172–3, 262
natural selection, 9, 66, 221, 230–1, 233–4,
    236, 255, 257, 259, 261–2, 270
NBS-LRR proteins, 27, 39–40, 97, 177
network, 190–7
nonself, 4, 5, 41, 49, 55, 99, 127–9, 267
Nossal, G J V, 81

Okasha, S, 3, 9, 221, 228, 235
organism, 10, 44, 238–9, 253–4
    heterogeneous, 242–62, 269
Owen, R D, 60–61, 72, 101
Oyama, S, 269

parasitism, 34, 122–5, 127, 144, 161, 168,
    174, 251, 254
Pasteur, L, 17
pathogen, 15, 17–8, 20, 26, 36–7, 39–40,
    93, 95, 124, 137, 163, 169, 176, 208,
    252–3
pathogen-associated molecular patterns
    (PAMPs), 25, 26, 39, 176, 211
pattern recognition receptor (PRR), 24,
    140–1, 164, 168
Pauling, P, 65
phagocytosis, 18, 23, 34, 50–1, 86–7, 94–5,
    139, 152, 163–5, 177
physiology, 12, 41, 55, 60, 75, 221, 230–1,
    233, 236–45, 250–1, 255–60, 262,
    263–5, 268–70
plant immunity, 25–6, 38–40, 95, 175–7

Queller, D C, 3, 9, 221, 228, 235, 251
Quine, W V O, 221

regulatory T cells, 35–6, 87, 90, 95–7,
    104–8, 112–5, 120–1, 125, 126, 150,
    154, 157–60, 165–70, 182, 202, 211–2
Reichenbach, H, 3, 249

RNA silencing, 27–9, 30–1, 176, 179, 253

Sakaguchi, S, 36, 96–7, 105
Sapp, J, 117
self, 4, 41–46, 49, 55, 57, 59, 69–70, 98–9,
    128–9, 191, 204, 219, 267
    genetic self, 44, 54, 58–60, 68–71, 78,
        82, 225
    as metaphor, 129
    self-nonself theory, 5, 55, 71, 77–81,
        128–30, 185–90, 193, 205, 214–8,
        226, 267
    see also: 'Burnet', 'identity', 'individual',
        'Metchnikoff', 'organism'
Silverstein, A M, 18, 46, 52, 65–6, 208–10
Snell, G D, 77
Sober, E, 3, 228–9, 235, 238, 251–2, 270
social immunity, 252–3
specificity, 15, 19–22, 25–6, 28–29, 31, 35,
        37–9, 65–6, 87, 89, 91, 135–6, 152–4,
        166, 169, 192, 211, 215, 225, 244
Sterelny, K, 235, 258
Strassmann, J E, 3, 9, 221, 228, 235
Strawson, P F, 3
substance, 3, 249
superorganism, 251–2
surveillance (immune), 88–9, 93–94, 172,
        176, 178, 188, 240, 246, 261–2, 264
symbiosis, 117–8, 150, 174, 247
    symbiotic bacteria: see 'tolerance'
systemic conception in immunology,
        190–205

T cells: see 'lymphocytes'
Talmage, D W, 65
Tauber, A I, 6, 18, 51, 57–8, 100, 129, 134,
        190, 193, 196, 209, 221
theories in immunology, 18, 64–68, 72,
        185–218

tolerance, 62–64, 67, 99–103, 126–8, 136,
        146–62, 174–5
    fetomaternal tolerance, 103, 108, 111–6,
        119, 159, 174, 206
    graft tolerance, 103–11
    'infectious' tolerance, 106
    self-tolerance, 60, 62, 215–8
    of symbiotic bacteria, 117–23, 150–1,
        155, 159–61, 167–8, 174, 181–2,
        186, 206, 247–8, 258–9, 262
toll-like receptor (TLR), 20, 29, 92–3, 122,
        139–40, 150, 160,164–5
Tonegawa, S, 79
transitions in evolution, 3, 261
transplantation, 5, 7–8, 19, 42–44, 53–55,
        60, 70, 72, 78, 101–11, 144, 157, 182,
        223, 241, 246
tumor: see 'cancer'

unicellular immunity, 29, 178–80, 263
uniqueness, 2–4, 6–10, 54, 78, 222–7, 268,
        270

vaccination: see 'immunization'
Varela, F J, 190, 196–9
virus, 17, 35–6, 58, 77, 89, 138, 161,
        208

Waldmann, H, 103, 105–6, 157
Weismann, A, 257, 260
Wells, H G, 42, 57, 58
Wiggins, D, 3, 249
Wilson, D S, 229, 251–2
Wilson, J, 3, 228, 232, 238
Wilson, R, 228, 232

Xu, J, 118, 121, 161, 248

Zinkernagel, R M, 79, 146,

CPSIA information can be obtained
at www.ICGtesting.com
Printed in the USA
BVHW072332131218
535589BV00008B/14/P